TI-82 Guide

for Moore's

The Basic Practice of Statistics

Larry Morgan
Montgomery County Community College

W. H. Freeman and Company
New York

To Abui, Esi, Narissa, and Kojo

ISBN: 0-7167-2923-7

Printed in the United States of America

First printing 1995, VB

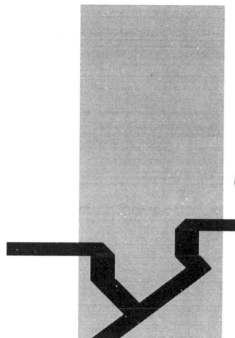

Contents

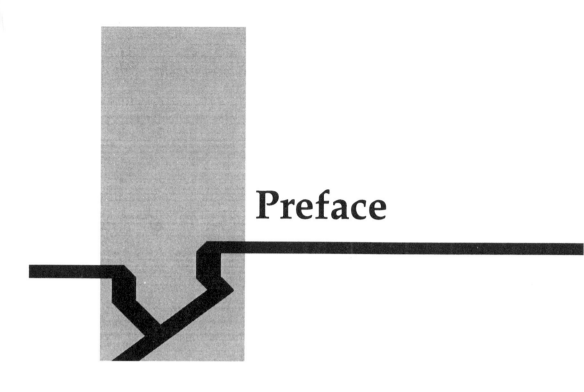

Preface

"The goal of statistics is to gain understanding from data. Data are numbers, but they are not 'just numbers.' Data are numbers with a context." [p. 3]

The TI-82 is a little computer with many capabilities to help you in your understanding of data and statistics. This guide will introduce you to those capabilities and explain how to use them as you study the text *The Basic Practice of Statistics* by David S. Moore. There is often more than one way of doing something on the TI-82, but this guide concentrates on the most valuable methods.

After a brief introduction, which includes an explanation of the TI-82 keyboard and the notation used throughout this guide, we follow the text chapter by chapter, providing helpful techniques for working with the TI-82. Most of the examples and data come from the text, with the text page number indicated in square brackets, as at the end of the quote above. Examples are paraphrased to make it easier to relate to the text, but the text should be referred to for important details, such as the appropriateness of each procedure.

The Introduction and Chapter 1 are very important for the understanding of the rest of the guide. Read them carefully and refer to them as often as necessary for definitions of such phrases as "home screen," the "last entry feature," and "running a program."

Many procedures are introduced that do not require programs but can be simplified with their use. Thus, appropriate programs are introduced as they are needed, and their output is related to the computer output in the text. These programs can be obtained from your instructor or from a disk available from the publisher. Also on the disk are all the data sets referenced in the text. Appendix A explains how these programs and data can be downloaded from a computer or another TI-82.

The Calculator Based Laboratory (CBL) system allows for dynamic gathering of data from a variety of sensors, such as temperature probes and force sensors. The data can be transformed from the CBL to the TI-82 for analysis. Appendix B provides two examples of how this equipment can be used with the statistics course.

Larry Morgan
Philadelphia, August 1995

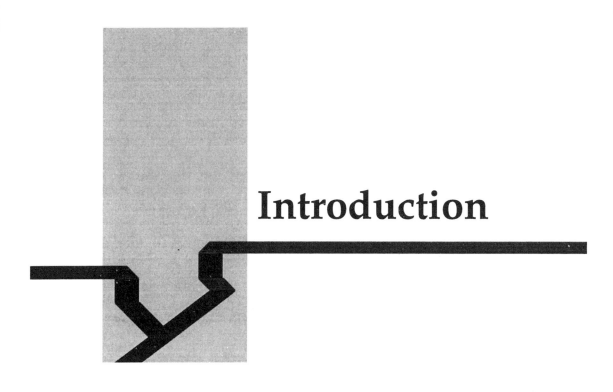

Introduction

The TI-82 Graphics Calculator is in reality a hand-held computer, which will be amply demonstrated by what we will do with it in our studies of statistics. This introduction will give you an overview of the TI-82 keyboard and the notation for the keys used in this guide.

In addition, we will show you the many possible settings for the **MODE** of the TI-82 in order to obtain the same results given in this guide.

We will also show you how to adjust the screen contrast to keep it sharp and bright and how to check the strength of your battery.

Also included is a section of important menus and the programs we will be using.

TI-82 KEYBOARD AND NOTATION

The TI-82 keyboard has five columns (designated A, B, C, D, E) and 10 rows of keys. The cursor control (or arrow) keys (▲ ▼ ◄ ►) toward the upper right of the keyboard disturb the pattern in a logical way (see the keyboard schematic on the following page).

Alphabetized Key Lookup Table		
Key	Loc.	Ex.
2nd	A2	p. 3
ALPHA	A3	p. 3
↟ **ANS**	D10	p. 19
CLEAR	E4	p. 10
DEL	C2	p. 11
ENTER	E10	p. 10
↟ **ENTRY**	E10	p. 10
GRAPH	E1	p. 25
↟ **INS**	C2	p. 11
↟ **LIST**	C3	p. 12
MATH	A4	p. 14
MATRX	B4	p. 32
MODE	B2	p. 4
ON	A10	p. 10
PRGM	C4	p. 6
↟ **QUIT**	B2	p. 19
STO▶	A9	p. 10
STAT	C3	p. 11
↟ **STAT/PLOT**	A1	p. 13
TRACE	D1	p. 14
VARS	D4	p. 26
WINDOW	B1	p. 14
Y=	A1	p. 26
ZOOM	C1	p. 17
÷ for /	E6	p. 11
× for *	E7	p. 12

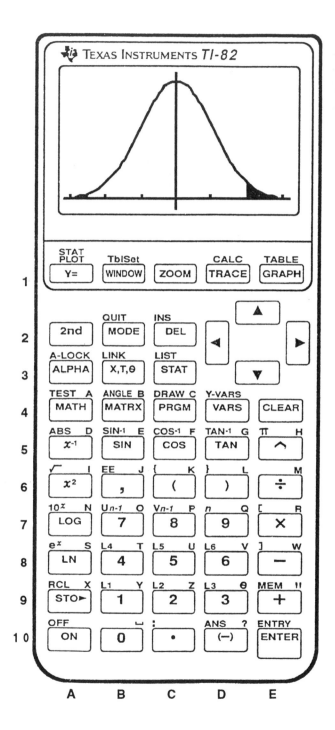

Touching the keys mentioned as you follow along, start in the upper left corner with the **Y=** key in the A column, row 1; or the A1 location. Find then the **ON** key at A10, the **ENTER** key at E10, and proceed up to the **GRAPH** key at E1, thus touching all four corners of keyboard.

Many keys on the TI-82 have multiple functions. The main function is marked on the key itself. Functions printed in blue above the key require you to first press the blue **2nd** key at the A2 location. The letters of the alphabet printed in gray above some keys are entered by first pressing the gray **ALPHA** key at the A3 location. Let's see how each of the alternative functions work.

Blue 2nd Key at A2

The key at the C3 location is marked **STAT**. Above it in blue is the word **LIST**. To use the **LIST** function, press and release the blue **2nd** key, then the **STAT** key. When you press the **2nd** key, the cursor turns into a blinking up arrow, ✦, which tells you that the next key you press will activate the function printed in blue above that key. We need a short way to show this, so in this manual, ✦ **LIST** means press and release the **2nd** key, then press the key with **LIST** above it in blue.

When we write ✦ **QUIT**, ✦ **INS**, ✦ **ANS**, or ✦ **ENTRY**, again hit the **2nd** key first. You must also use the **2nd** key for the list designations **L1** to **L6** that appear in blue above the number keys **1** to **6**, and for the braces { } that appear in blue above the parentheses ( ) keys at C6 and D6. We will not use the ✦ notation for these, because their locations are easy to remember.

Gray ALPHA Key at A3

The letters of the alphabet can be used to designate data storage locations and are engaged by first pressing and releasing the **ALPHA** key. For example, the letter **F** can be engaged by hitting the gray **ALPHA** key then the **COS** key at C5 that has the letter **F** above it in gray. When you press the **ALPHA** key, the cursor turns into a blinking A (for Alpha), which tells you that the next key you press will duplicate the letter printed in gray above that key onto the TI-82 screen.

Some General Keyboard Patterns and Important Keys

1. Row 1 is for plotting and graphing.

2. Row 2 has the important **2nd** ↟ **QUIT** combination and the **DEL** ↟ **INS** keys. The cursor control keys, which are used for editing, are on the right. Just below the cursor control keys is the **CLEAR** key.

3. The A column has the **MATH** key and math functions, e.g. x^2, $\sqrt{}$.

4. The E column has the math operations **+, – , ×, ÷, ^** (Note that ÷ shows on the TI-82 screen as **/** and × as **∗.**)

5. The **STAT** ↟ **LIST** key at C3 sits above three important keys in row 4, the **MATRX, PRGM,** and **VARS** keys.

6. Row 6, has the **({ , })** keys, which are used for grouping and spacing.

7. The **STO▶** key at A9 is used for storing and shows as **➡** on the display screen.

8. Row 10 has the negative symbol **(-)** ↟ **ANS** key, which differs from the subtraction key in the E column, and the biggest key, the **ENTER** ↟ **ENTRY** key. (Note that **(-)** shows as ⁻ on the screen, smaller and higher than the subtraction sign.)

SETTING THE CORRECT MODE: THE MODE KEY AT B2

If you do not get as many decimals in your answers as in the examples in this guide or have difficulty getting other output, check your mode settings.

From the home screen, the screen that appears when you turn on your T1-82, press **MODE** and the menu on the right appears with the first word in each row darkened. If your screen looks different use the cursor control key (▼) to go to each row and with the first element blinking press **ENTER**. Repeat this procedure

until the screen looks as we want it. Try this and all other examples on your TI-82 for better retention.

Engage ↟ QUIT at B2 to return to the home screen.

TI-82 SCREEN CONTRAST ADJUSTMENT AND BATTERY CHECK

From the home screen, press and release the **2nd** key and hold down the ▲ cursor control key to increase the contrast. Notice the number in the upper right corner of the screen increases from a possible low of zero to a maximum value of 9 as you hold down the ▲ key.

Press and release the **2nd** key and hold down the ▼ key to decrease the contrast. (If you adjust the contrast setting to zero, the display may become completely blank. Press and release the **2nd** key and then hold down the ▲ key until the display reappears.)

When the batteries are low, the display begins to dim (especially during calculations), and you must adjust to a higher contrast setting. If you find it necessary to set the contrast setting at 8 or 9, you need to replace four AAA batteries soon. (The display contrast may appear very dark after you change batteries. Press and release the **2nd** key then hold down the ▼ key to lighten the display.)

IMPORTANT MENUS

The keys that call up these menus are given first followed by a page or pages where you will find that menu used in this guide.

STAT (pp. 11 + 18)   ↟ **LIST** (pp. 38 + 12)

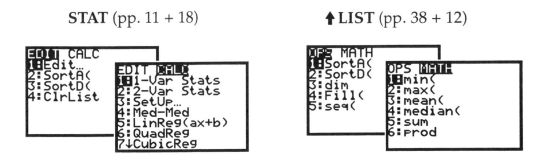

MATH (pp. 14 + 38)

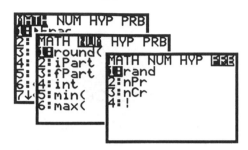

VARS 5: Statistics (p. 26)

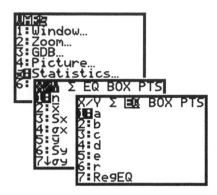

♠ STAT PLOT (p. 13)

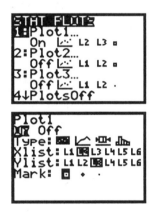

WINDOW (p. 14)

ZOOM (p. 17)

THE PROGRAMS OF MSTATPAK

When the MSTATPAK.82G file is transferred from the computer to a TI-82, its programs are rearranged into the individual programs listed on the right. They are called up by pressing the **PRGM** key and the ▼ key to scroll down the listing. Note that the program names begin with the letter A followed by numbers 1 to 8 and more letters. Program PDIST is listed as A1PDIST, which we will indicate in this guide as A1**PDIST**.

A brief description of each program follows, including where in this guide you will find an example of how to use it. See Chapter 1 (page 20) on how to run programs in general.

A1**PDIST** has seven options, as shown in the screen on the right.

    1. Z-TO-P calculates the area under a standard normal curve between two inputted values Z1 and Z2, with the option of plotting and shading the desired area (page 21).

    2. P-TO-Z, which requires a left tail area of a standard normal curve, calculates the output of the corresponding Z value approximately (page 22).

The next three options give the probabilities or areas in both tails for the given distributions for the appropriate input. The F-DIST option gives only the right tail probability. The Z-DIST is accurate to the four places given. The other distributions are sometimes approximate but work well over a wide range. Examples of these options are found in this guide as follows:

    3. Z-DIST (page 21)

    4. T-DIST (page 65)

    5. CHISQ-DIST (page 79)

    6. F-DIST (page 69)

A2**BINOM** gives binomial probabilities and sums of probabilities from inputted N, P, and the range of values of interest. An option is a plot of the distribution and a normal curve fitted to the distribution (page 48).

A3**CHART** gives control charts for time values in L1 and data in L2 and input of LCL, center line, and UCL (page 53).

A4LINREG does linear regression analysis, for X values in L1 and Y in L2, which includes scatterplot, regression line, correlation, hypothesis testing, predictions and residual plot (page 27).

A5TSAMPS does two independent sample calculations for hypothesis testing and confidence intervals: 1:MEANS (page 67) and 2:PROPORTIONS (page 73).

A6TWTAB gives two-way table analysis for data or observed frequencies in Matrix [D]. Percentages by cell, row, column, and margins are given, in addition to chi-square analysis (page 32).

A7RANDOM gives the following four options using a random number generator:

> 1. Generate random counting numbers that can be used for taking random samples (page 39).
>
> 2. Simulate the tossing of coins (page 45).
>
> 3. Simulate the Central Limit Theorem by tossing 1, 2, 4 or 8 nine-sided dice (page 51).
>
> 4. Simulate sampling a population with a population proportion of an inputted P (page 42).

A8ANOVA has two options but only the first, for one-way analysis of variance, will be used in this guide. 1:ONE-WAY uses input of means, standard deviations, and sample sizes or with data in Matrix [D] (page 83).

AA0UTILS is a utility program for swapping columns of Matrix [D] with list L6 and a program to calculate the mean, standard deviation, and sample size for one variable data stored by rows in Matrix [D] (page 55).

CHAPTER 1

Examining Distributions

We begin our study of statistics using *The Basic Practice of Statistics* and the TI-82 by taking an example from the text and showing how to do basic calculations and editing on the home screen of the TI-82. The home screen has the advantage of keeping past entries in sight so that future steps follow more easily.

Next we show the spreadsheet capabilities of the TI-82 for storing data and doing calculations. We then draw a number of important statistical plots and complete the chapter by running programs that eliminate the need to look up some values in the text tables.

Be sure to read the material in "Introduction to the TI-82," particularly the sections on key location and notation. It is suggested that you follow along with each example by keying them into your own TI-82. The useful information in this chapter on statistics and the TI-82 will serve you well throughout your course and throughout this guide.

HOME SCREEN CALCULATIONS

Categorical Variables

Example [p. 13]   The distribution of marital status for all Americans age 18 and older are given in the table below. We see that the largest percentage of people are married (61.1%). Based on the counts, we will check the calculations of the percents on the home screen of the TI-82.

Marital status	Count (millions)	Percent
Single	41.8	22.6
Married	113.3	61.1
Widowed	13.9	7.5
Divorced	16.3	8.8

1. When you press **ON**, your TI-82 is on the home screen with a flashing cursor. If there is previous work on the screen it can be removed by pressing the **CLEAR** key at E4. The **CLEAR** key removes a line to the left or the screen above.

2. Type 41.8+113.3+13.9+16.3, then press **ENTER**, for a total of 185.3 as shown in the first three lines of the screen display on the right.

```
41.8+113.3+13.9+
16.3
                185.3
Ans→A
                185.3
41.8+113.3+13.9+
16.3→A■
```

3. Store this sum as **A**. Press **STO►** at A9 and Ans→ appears on the screen. Engage **A** by pressing **ALPHA MATH** then **ENTER** for 185.3.

This procedure can also be done in one step, as in the last two lines of the display, using the last entry feature of the TI-82.

Engage **↑ENTRY** (or **2nd ENTER**) for Ans→A and again **↑ENTRY** for the first line of the recalled output. Press **STO►A** for the last lines in the previous screen display and the first lines of the display on the right. **ENTER** gives the total 185.3 once again, but this time stored as A.

```
41.8+113.3+13.9+
16.3→A
                185.3
41.8/A
        .2255801403
113.3/A
        .6114409066
13.9■/A
```

4. To calculate percentages: For the fraction of our data that are single, type 41.8 then ÷ then **A** for 41.8/A on the screen. Press **ENTER** to get 0.22558 or about 22.6%.

Press ↑ **ENTRY** to bring back the previous line. Use the ▲ cursor control key to jump to the front of the line. Type 113. over the 41.8, then press ↑ **INS** (or **2nd DEL** at C2) to change the flashing rectangular cursor to the flashing underline of the insert mode. Type 3 to get 113.3/A on the screen. **ENTER** gives 0.61144 or about 61.1% married.

Press ↑ **ENTRY** to bring back the previous line. Use the ▲ cursor control key to jump to the front of the line. Type 13.9 over 113. (see the last line of the display screen with the flashing black rectangular cursor over the final 3 of the previous 113.3), then press **DEL** to delete the 3. **ENTER** gives 0.0750 (not shown) or 7.5% widowed. This procedure can be continued for the 8.8% divorced with a minimum amount of keying.

SPREADSHEET CALCULATIONS

The TI-82 works well as a calculator with the home screen showing past steps and being able to recall past entries but there is often a better way of doing statistical calculations using the Spreadsheet capabilities of the TI-82.

Example   (Continue with the same example from [p. 13])

1. To clear lists or columns for data: Press **STAT** to get the screen on the right, then **4** followed by **L1,L2,L3** (above 1, 2, and 3 – don't forget the comma key at B6), then **ENTER** for "Done," as in the second screen, which indicates L1, L2, and L3 are cleared of all values. From now on such a multistep procedure will be shown as **STAT 4:** ClrList **L1, L2, L3 ENTER.**

Note: If you forget to clear lists before entering the spreadsheet, see step 5 for another way to clear lists from within the spreadsheet.

2. To enter the data into the spreadsheet: Press **STAT 1:** Edit, then type 41.8. Press **ENTER**, then type 113.3 and the spreadsheet will look like that on the right. Press **ENTER** and the 113.3 at the bottom of the screen is posted in the second row and the cursor moves down to the third row. Continue with 16.3 **ENTER**, then 13.9 **ENTER.** Note that the data was entered with the last two values interchanged to show the edit features below.

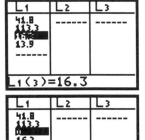

3. To edit a spreadsheet: To insert a value above 16.3, move the cursor to 16.3 as in the screen on the right and engage ⬆ **INS**. A zero appears above 16.3 as in the bottom screen. Type 13.9 then **ENTER** to insert this value in the proper place.

To delete a value move the cursor to that value, the bottom 13.9 in this case, and press **DEL.** The value is deleted. If there were data below, they would move up a space.

4. To calculate percentages: Highlight L2 as on the screen on the right, then type **L1** ÷ ⬆ **LIST** [MATH] **5:sum L1**, and the bottom line will look like that on the right. Note: [MATH] is a menu of ⬆ **LIST** (at C3) obtained by moving the cursor key to highlight it. Pressing **5** has the fifth function of this menu, "sum," posted on the screen. (See Important Menus in "Introduction to the TI-82.")

Press **ENTER** and each value of L1 is divided by its sum as in the screen on the right. Do a similar procedure letting L3 = **L2** x 100 (for L2*100 on the screen) if you wish to change from decimal to percent.

5. To clear a list within the spreadsheet: Move the cursor up until L2 is highlighted. Press **CLEAR** and the bottom line is cleared but not the column, as in the screen on the right. Press **ENTER** and the column is also gone.

DRAWING HISTOGRAMS

From Grouped Data

Example [p. 16]   A count of states with percentages of their population over 65 years of age is duplicated from the text in the table below with added columns for class marks, the arithmetic average of the two end limits (e.g., 5.5 = (5.0 + 6.0)/2).

Class			Class mark	Count	Class			Class mark	Count
4.1	to	5.0	4.5	1	12.1	to	13.0	12.5	10
5.1	to	6.0	5.5	0	13.1	to	14.0	13.5	12
6.1	to	7.0	6.5	0	14.1	to	15.0	14.5	5
7.1	to	8.0	7.5	0	15.1	to	16.0	15.5	4
8.1	to	9.0	8.5	1	16.1	to	17.0	16.5	0
9.1	to	10.0	9.5	0	17.1	to	18.0	17.5	0
10.1	to	11.0	10.5	9	18.1	to	19.0	18.5	1
11.1	to	12.0	11.5	7					

Note: Before we draw any statistical plots, it is important that any function that shows when the **Y=** key is pressed be cleared or turned off.

1. Clear the data from L1, L2, and L3 and then enter 15 class marks in L1 and the counts in L2 so that your spreadsheet looks like the screen display on the right.

2. Engage ↑ **STAT/PLOT** (above **Y=** at A1), which gives something like the screen on the lower right. Because several screens are on, press 4: PlotsOff **ENTER** for "Done," to turn off all plots. (Of course, if all your plots are already off, this step is not needed.)

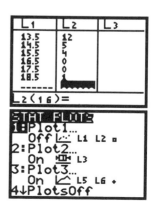

3. Engage ↑ **STAT/PLOT** again, but this time press **1**: Plot1. A screen like the one below left appears, but if yours looks different, read on. Use the ◀ cursor control key to highlight ON, then press **ENTER** to activate. Use the cursor control keys to highlight other changes and press **ENTER** after each change. Notice the "Type" is the last on the right, the "Xlist" values are in L1 and the "Freq" (or count) is in L2 (see the screen below right).

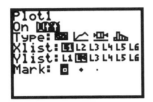   

4. Press **WINDOW** (at B1) and then **ENTER**. Press **4 ENTER** for "Xmin=4," like the screen on the right. Continue with "Xmax=19" and "Xscl=1" or the width of each class (e.g., 4 to 5, 5 to 6, etc.). The maximum count was 12, so we set "Ymax=13" to include 12.

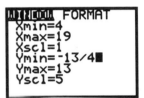

"Ymin=⁻13/4" or one-fourth of Ymax but negative (be sure to use (⁻) in the last row for the negative sign).

When you press **ENTER** this will change to ⁻3.25 but leaves room at the bottom of the histogram display to show the cell boundaries along with the count for a class (as shown in the next step and display). "Yscl" gives the distance between tic marks on the Y axis.

5. Press **TRACE** (at D1) and use the ▶ key to find the counts for each class, as shown in the screen on the right and the screen below. Notice that there are seven states (n = 7) that have more than 11% but at most 12% of their population 65 years of age or older.

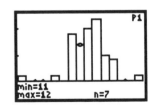

Note: To plot histograms for grouped data the "Freq" values must be integers. If you are using percentages, as in problem 1.6 of the text [p. 23], round these values to the nearest percent (zero decimals) with something like L2 = round(**L2, 0**) with "round" under **MATH** [NUM].

From Raw (Ungrouped) Data (99 values or less)

Note: We will store the following data in a list that has a limit of 99 values. If you have more data than that, see the last pages of Chapter 4.

Example (Table 1.1 of the text [p. 15])
The data, the source of the above grouped example, is:

12.9, 4.2, 13.2, 14.9, 10.5, 10.1, 13.7, 12.2, 18.3, 10.1, 11.4, 12.0, 12.5, 12.6, 15.4, 13.9, 12.7, 11.2, 13.4, 10.9, 13.7, 12.1, 12.5, 12.4, 14.1, 13.4, 14.1, 10.8, 11.6, 13.4, 10.9, 13.1, 12.3, 14.5, 13.1, 13.5, 13.7, 15.5, 15.1, 11.4, 14.7, 12.7, 10.1, 8.8, 11.9, 10.9, 11.8, 15.1, 13.3, 10.6

1. Enter the data into L1. Change Plot 1 under ↑ **STAT/PLOT** in order to have Freq:1, because you are going to count each of the 50 values one at a time (see display at right).

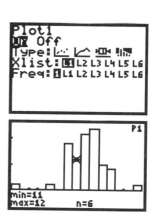

The window remains the same as for the grouped data, so when you press **TRACE** and use the ► key, the histogram on the right appears. Note that this differs a bit from the previous histogram because the TI-82 puts the 12.0 value of Idaho in the 12 to 13 class instead of the 11 to 12 class. Thus, the 11 to 12 class has one less than before and the 12 to 13 class one more, but the general appearance of the histograms are the same.

2. The importance of Xscl: If under **WINDOW** Xscl is changed to Xscl = 2, and **TRACE** is then pressed, you get the histogram shown in the first display screen at the top of the following page. Using ► we see a maximum count of 23 in the 12 to 14 class and change Ymin and Ymax accordingly, as shown in the second screen (with Ymin = ⁻24/4). Press **TRACE** to see the histogram in the third screen, with fewer but wider cells. Note: It is not difficult to set up Xmin and Xmax but Ymax is not so easy. The point is to try something and then correct it with the use of the **TRACE** key. (Set Xmax=20 then **GRAPH** or **TRACE** to display all of the last cell that goes from 18 to 20.)

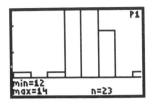 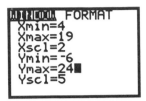 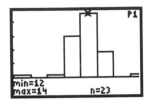

Making a Duplicate List

From the home screen, with **L1 STO▶ L6 ENTER** we can save a copy of L1 in L6 for later use.

STEMPLOTS

Example [p. 24]　Data as in the previous example saved in L6 (percentage of the population over 65 years old by state).

The TI-82 will put the data in order but first make a duplicate of the data in case the original order, alphabetically by state, will be needed.

1. Type **L6 STO▶ L1** then **L6 STO▶ L2**.

2. Putting a List in Order: Press **STAT 2:** SortA(then **L2** and **ENTER** or **STAT 2:** SortA(**L2 ENTER** for "Done."

3. To check on the above list, press **STAT 1:** Edit and you will see the original data in L1 in ascending order in L2 starting with the low value of 4.2 as shown in the display on the near right. By going down this list one can more easily build a stemplot, part of which is shown at the far right and all of which is in the text.

L1	L2			
12.9	4.2		4	2
4.2	8.8		5	
13.2	10.1		6	
14.9	10.1		7	
10.5	10.1		8	8
10.1	10.5		9	
13.7	10.6		10	11156

L2(1)=4.2

TIME PLOTS

Example [p. 27]　Death rate from cancer (death per 100,000 people).

Year	1940	1945	1950	1655	1960	1965	1970	1975	1980	1985	1990
Death	120.3	134	139.8	146.5	149.2	153.5	162.8	169.7	183.9	193.3	201.7

1. Put the year data in L1 and deaths in L2.

2. Be sure all plots are off but one with ↑ **STAT/PLOT**, like the one on the right, with the years in the Xlist and deaths in the Ylist.

3. Press **ZOOM 9:** ZoomStat then **TRACE** and ▶ to get the following screen, which is similar to Figure 1.7 in the text.

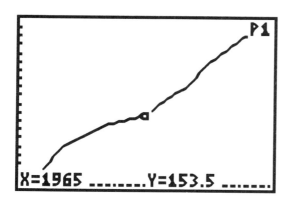

DESCRIBING DISTRIBUTIONS WITH NUMBERS WITH 1-VAR STATS

Example [p. 37]   A study examined the number of hysterectomies performed in a year by doctors.

1. Put the number of operations by a sample of 15 male doctors in L1: 27, 50, 33, 25, 86, 25, 85, 31, 37, 44, 20, 36, 59, 34, 28.

2. Press **STAT** [CALC] **1:** 1-VarStats **L1 ENTER** to get the screen at the top of the following page.

(a) The Mean = $\bar{X}$ =41.333 [Example 1.7, p. 37]. We will discuss other values on this screen later but notice the down arrow in the last line. This can be reached by using the ▼ cursor control key for the next screen down with:

(b) The **Median** Med=34 [Example 1.8, p. 39]

(c) The **Range** = maxX–minX=86–20 =66 [Example 1.9, p. 41]

(d) **Quartiles** Q1=27 and Q3=50 [Example 1.10, p. 42]

(e) **Five-Number Summary** minX= 20, Q1=27, Med=34, Q3=50, maxX=86 [Example 1.10, p. 44]

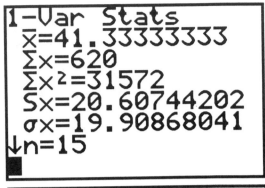

BOXPLOTS

Example [pp. 43-44]

1. With the male doctor data in the above example in L1, put the following female doctor data in L2: 5, 7, 10, 14, 18, 19, 25, 29, 31, 33.

2. Under ↑ **STAT/PLOT** set up the two screens below at left.

3. **ZOOM 9** then **TRACE** to get the screen on the right below with the median for P1, the male doctor data, highlighted and displayed as Med=34. The ◄ ► keys reveal all of the five-number summary values; ▲ or ▼ switches to P2.

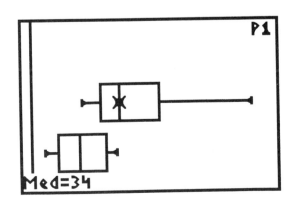

Note: If the axes are in the way, change in **WINDOW:** Ymin=1, Ymax=2, and Yscl=0 or use AxesOff under **WINDOW** [FORMAT] but then remember to later change to AxesOn. Note that a third plot would fit on the screen if you had a third set of data.

STANDARD DEVIATION

Example [p. 47]   Metabolic rates of 7 men who took part in a study of dieting follows (the units are calories per 24 hours): 1792, 1666, 1362, 1614, 1460, 1867, 1439

1. **Using 1-VarStats:**
(a) Put calorie data in L1.
(b) Press **STAT** [CALC] **1:** 1-VarStat **L1 ENTER** for Sx= 189.2397069, the fourth value of the output on the right.

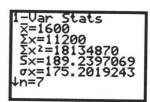

2. **Using spreadsheet and definition:**
$$s^2 = (\Sigma(x-\bar{x})^2/(n-1)$$
(a) With the calorie data in L1, under **STAT 1:** Edit, highlight L2 and enter **L1** - mean (**L1**, with "mean" under ↟ **LIST**[MATH]) to get the first screen below. Press **ENTER** for the differences in the second screen.
(b) Highlight **L3** and enter L2², as in the second screen, then **ENTER** for the third screen with the differences squared. (Note that the X^2 key is at A6
(c) Engage ↟ **QUIT** to return to the home screen.

L1	L2	L3
1792	------	------
1666		
1362		
1614		
1460		
1867		
1439		

L2=L1-mean(L1▮

L1	L2	L3
1792	192	------
1666	66	
1362	-238	
1614	14	
1460	-140	
1867	267	
1439	-161	

L3=L2²▮

L1	L2	L3
1792	192	36864
1666	66	4356
1362	-238	56644
1614	14	196
1460	-140	19600
1867	267	71289
1439	-161	25921

L3(1)=36864

(d) Enter sum L3/6, with "sum" under ↟ **LIST**[MATH], for the variance s^2= 35811.66667 as on the screen on the right.

```
sum L3/6
         35811.66667
√Ans
         189.2397069
```

(e) √ at A6 followed by ↟ **ANS** (at D10) then **ENTER** for the standard deviation of 189.2397069.

Note: A difficulty of using ↑ **LIST** for calculations with 1-VarStats is that we are limited to 99 values. If you have more than 99 values, see the last pages of Chapter 4.

NORMAL DISTRIBUTION CALCULATIONS PROGRAMS

This is your first program. If you should wish to leave a program before given the option to QUIT, just press **ON** and the screen on the right will appear.

Press **2**:Quit to return to the home screen. Be careful! If you do anything else, you may go to a line in your program editor. If this should happen, press ↑ **QUIT** at B2 to return to the home screen. If you do otherwise, it could change your program and it will have to be reloaded.

Note: If your programs were protected when first loaded from a computer to a TI-82 using Graph Link for Windows or the comparable software for Macintoshes, you will only be given the Quit option and will never have this problem.

Proportions from Z Values

Example [p. 66]   What is the area under the normal curve less than (to the left of) $z = 1.4$?

1. Run program A1**PDIST** 3:Z-DIST (steps a, b, c).

Note: The A1 in this program name is designed to have it appear at the top of a group of programs.

(a) Press **PRGM** (at C4) for the screen on the right.
(b) Press **1** for prgmA1PDIST to appear on the home screen. (If your program does not appear first on the list then press the appropriate number or highlight the program using the cursor key and press **ENTER**.)

(c) Press **ENTER** for the screen on the right. Press **3** for Z-DIST and the second screen.
(d) At the Z=? prompt, enter 1.4, then press **ENTER** for an area of 0.9192, as shown in the third screen.

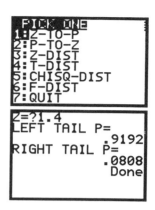

Note: In the rest of this guide, steps like a, b, and c will be given as "Run program A1**PDIST** 3:Z-DIST."

Notice the pause indicator (moving dots) in the upper right corner of the first screen. This indicator alerts you that the TI-82 is waiting for some input or an **ENTER** to continue. This changes to the busy signal (moving bar) as the calculations are being performed in the second screen.

2. Run program A1**PDIST** 1:Z-TO-P
We could have done the previous example using this program, but this program is looking for two Z values as input.

(a) Enter ‾4 at the Z1 prompt because there is no point going further to the left. It will not effect the area under the curve to the 4th decimal place.
(b) Enter 1.4 for Z2 then **ENTER** for 0.9192 as before.
(c) Press **ENTER** again and the menu in the second screen on the right is displayed. Press **2** for the PLOT option that gives the figure with the area, given above, shaded.
(d) Press **ENTER** and **3** to QUIT.

This program is of great advantage if you want to calculate the area between two z values or if a plot is desired. If you are only interested in the area of a tail, then the first program is the one to use.

Finding a Z Value Given a Proportion

Example [p. 71]   To find a z-value that goes with the top 10%.

1. Run Program A1**PTAILS 2:**P-TO-Z

2. When the prompt P=? appears, we are looking for the proportion to the left of the desired Z value, so type 1-0.1 or 0.90 and **ENTER** for Z=1.28, as in the screen on the right and in the text.

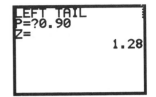

Note that this program also has the plot option.

Note also that this program gives good approximations most of the time but is not as accurate as the previous two programs, which are accurate to the four decimal places given.

CHAPTER 2

Examining Relationships

This chapter examines relationships between two quantitative variables. When looking for linear or straight-line relationships, we will be aided by the program A4**LINREG**. We will also investigate relationships between categorical data in two-way tables with the assistance of program A6**TWTAB**.

INTRODUCTION

Example [p. 92]   We will start examining the relationship between two quantitative variables: percent of a state's high school seniors who take the SAT and their median math score for each state. We will then introduce the categorical variable region or location of the state. We will only use the first 17 states listed in the text and on the right, to simplify the instructions.

State	Region	L1 Percent	L2 Math
AL	ESC	8	514
AK	PAC	42	476
AZ	MTN	25	497
AR	WSC	6	511
CA	PAC	45	484
CO	MTN	28	513
CT	NE	74	471
DE	SA	58	470
DC	SA	68	441
FL	SA	44	466
GA	SA	57	443
HI	PAC	52	481
ID	MTN	17	502
IL	ENC	16	528
IN	ENC	54	459
IA	WNC	5	577
KS	WNC	10	548

Scatterplots [p. 96]

1. Enter the percentages into L1 and the math scores into L2. (Preserve the order so that the points plotted for each state have the proper X and Y coordinates.)

2. With all other plots off, define Plot 1 as shown on the right.

3. Press **ZOOM** 9:ZoomStat then **TRACE** for the plot shown on the right.

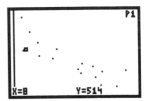

4. Because the information at the bottom of the screen is too close to the plotted points, use **WINDOW** to change Ymin=400 from 427.6 as below on the left.

5. Press **TRACE** and the ▶ cursor control key to identify the smallest Y values as in the plot below on the right.

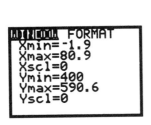
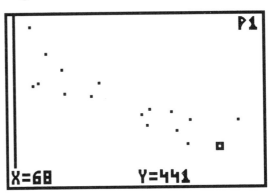

Checking with the table, we see the two lowest values are DC and GA, both in the SA (South Atlantic) region. We also notice in the plot two clusters of points. The cluster on the left, the states that have the smaller percent of students taking the SAT, has the higher median SAT math scores [p. 98]. To check if there is a pattern to the other states in the SA region, we plot them with another symbol.

6. Store 58, 68, 44, and 57 in L3 and 470, 441, 466, 443 in L4 (these values are boxed in the middle of the table on the first page of this chapter).

7. Define Plot 2 as shown below on the left and leave Plot 1 on as before.

8. Press **GRAPH** for the plot below on the right. Notice that the four SA states of first seventeen investigated all lie in the lower right cluster. The text works with all 51 values and increases the regions of interest [p. 103]. You might want to do your own investigations.

Linear Relationships

Example [p. 101]   Obtain the linear relationship shown in the scatterplot below of the average amount of natural gas used per day by the Sanchez household in 16 months against the average number of heating degree-days in those months from the table on the right.

Month	Deg-(L1) days	Gas (L2) (100 cu ft)
Nov.	24	6.3
Dec.	51	10.9
Jan.	43	8.9
Feb.	33	7.5
Mar.	26	5.3
Apr.	13	4
May	4	1.7
June	0	1.2
July	0	1.2
Aug.	1	1.2
Sept.	6	2.1
Oct.	12	3.1
Nov.	30	6.4
Dec.	32	7.2
Jan.	52	11
Feb.	30	6.9

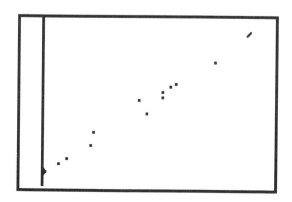

1. Store the degree-day values in L1 and the gas consumption in L2. Save this data for the next few sections.

2. With Plot 1 defined as it was in the previous example (all other plots are off), press **ZOOM 9**:StatPlot for the plot above.

CORRELATION

Example [p. 115]   To calculate the correlation coefficient between degree-days and gas consumption for the previous example, press **STAT**[CALC] 9:LinReg(a+bx) **L1**, **L2 ENTER** for the screen on the right, with r=0.9953 and as in the text.

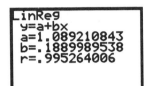

Note: Exercise 2.17 is worked out in detail at the end of this chapter, but try it now before checking it out there.

THE LEAST-SQUARES REGRESSION LINE

We see from the last screen display that the regression line for that data is
$\hat{y} = a + bx = 1.0892 + 0.1890x$, which agrees with the values in the text [p. 123].

Plotting the Regression Line (p. 120)

This procedure must follow that for the correlation coefficient so that the correct regression equation is stored in the TI-82.

1. After using **STAT**[CALC] 9:LinReg(a+bx) **L1**, **L2 ENTER**, press the **Y=** key (at A1). With the cursor next to a cleared Y1 (as in the top left screen on the following page), press **VARS 5**:Statistics...▶▶[EQ]7:RegEQ. The regression equation is pasted by Y1 as shown on the lower left screen.

2. If the STAT PLOT is defined as before, press **ZOOM:9**ZoomStat and then **TRACE**, which gives the screen shown at right on the following page. Note by using ▲ and ▶, we traced on the regression line to a point close to X=20 and YHAT=4.874.

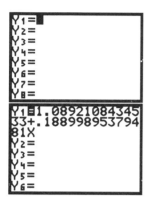
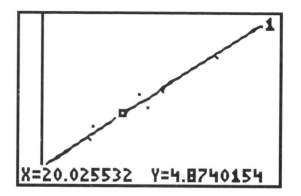

Predicting a Value Using the Regression Equation

To predict Y for X = 20 as in the text [p. 123], we can plug 20 in for X in the approximate regression equation as in the first lines of the screen on the right, or enter Y1(20) with Y1 under ↑ **Y-VARS** 1:Function...1:Y1, or use the table feature of the TI-82.

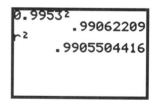

Calculating r² in Regression

r^2 can be calculated on the home screen as on the right, or to carry all decimal places r can be pasted to the home screen and squared. Press **VARS** 5:Statistics...▶▶ [EQ] 6:r, and it will be pasted to the home screen. Press the x^2 key (at A6), then **ENTER** for 0.9905504416 or 0.9906 as in the text [p. 127].

All that has been done with the gas consumption data in the last few pages can be automated with the following program.

PROGRAM A4LINREG

1. Running the program A4**LINREG** gives us the screen on the right, which reminds us that the input data must be in L1 and L2. This is what we have in the above example, but if this were not the case, after pressing **ENTER** the next screen permits us to quit by pressing 2.

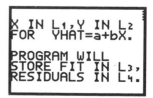

2. When we continue by pressing **1** we obtain the option screen.

3. Press **1** for the scatterplot and regression line displayed in the right screen below.

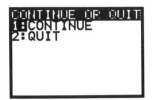 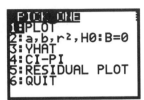 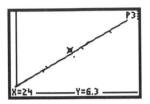

4. Pressing **ENTER** returns us to the selection screen and **2** brings the screen on the right, which gives us the a and b of the regression equation YHAT=a+bX, the correlation coefficient r and r^2, all with the same values as before. The rest of this screen will be discussed in Chapter 10.

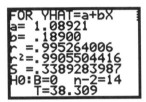

5. Pressing **ENTER** and **4** lets us input 20 for X to predict a Y value of 4.869 as in the screen on the right.

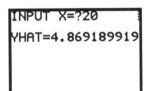

6. Press **ENTER** and the option screen returns. Press **6** to quit at this time. The next example will go over another option.

RESIDUALS

Example [p. 129]   Does the age (in months) at which a child begins to talk predict the later score on a test of mental ability? See the data on the right at the top of the following page.

1. Store the age in L1 and the score in L2.

2. Run program A4**LINREG** and note that the first screen indicates that the program will put the fit (YHAT) in L3 and the residuals in L4.

3. Continue to the option screen and pick **1:PLOT**, then use the ▶ cursor control key to highlight the outlier as in the plot below, with X=17 months and a score of 121.

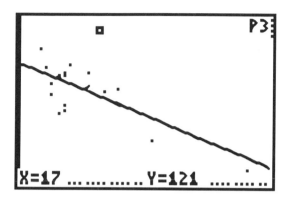

4. Press **ENTER** then select **5** for the residual plot below after the cursor key was used to highlight the outlier (child 19) who first spoke in the 17th month (with a score of 121). We see the residual is approximately 30.285.

Child	Age(mos.) L1	Score L2
1	15	95
2	26	71
3	10	83
4	9	91
5	15	102
6	20	87
7	18	93
8	11	100
9	8	104
10	20	94
11	7	113
12	9	96
13	10	83
14	11	84
15	11	102
16	10	100
17	12	105
18	42	57
19	17	121
20	11	86
21	10	100

5. Press **ENTER** then select **6:QUIT** and **STAT 1:**Edit. Use the cursor control keys to obtain the above residual as highlighted below right. The values shown in L4 are the values in the third row of residual in the text [p. 131].

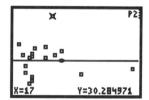

L2	L3	L4
102	97.477	4.523
100	98.604	1.396
105	96.35	8.65
57	62.54	-5.54
121	90.715	30.285
86	97.477	-11.48
100	98.604	1.396

L4(19)=30.28497...

INFLUENTIAL OBSERVATIONS

1. Using the same data as above, rerun program A4**LINREG**. With the ▶ cursor control key, obtain the plot with the child 18 highlighted, as on the right, who first spoke at 42 months and had a score of only 57. Is this an influential observation?

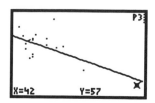

2. Press **ENTER** and select **2** for the screen on the right. Notice that the values of a and b of the regression line and the value of r^2 equals about 41%.

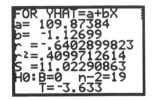

3. Press **ENTER** and select **6:QUIT**, then press **WINDOW** and note the values in the screen below on the left.

4. In order to save the window values, press **ZOOM**[MEMORY] **2:ZoomSto** from the menu, shown below on the right. The plot will reappear.

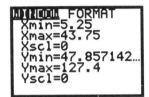

5. Press the **Y−** key and type or paste the regression equation so the screen looks like the one on the right.

6. Under **STAT 1:Edit...** delete the data for child 18 in L1 and L2, then rerun the program and plot for the results below on the left. Notice that the dimensions of this window are different than that in the previous plot.

7. Press **ENTER** and select **2** for the screen below on the right. Notice that r^2 drops to only about 11.2% [p. 135].

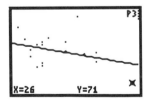

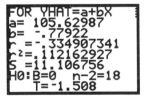

8. Press **ENTER 6:QUIT** then **ZOOM** [MEMORY] **3:ZoomRcl**, and the plot is redrawn with the original window.

9. Press **Y=** then ◄ to move the flashing cursor over the = sign. Press **ENTER** so the screen looks like it did in step 5 above. This changes the plotOff to plotOn. It was automatically turned off when we ran the program.

10. Press **GRAPH** (at E1) for the plot below and as in the text [p. 135]. Note the change in the slope of the line. What do you think would happen if we removed the highlighted point (obtained with **TRACE** and the ► cursor control key)?

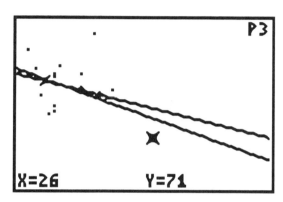

RELATIONS IN CATEGORICAL DATA

Example [p. 151]   The following table presents Census Bureau data on the years of school completed by Americans of different ages (thousands of persons).

Education	Age group				
	25-34	35-44	45-54	55-64	≥65
Did not complete h.s.	5965	4755	4829	5999	12702
Completed high school	17505	14498	10300	8645	10310
1-3 years of college	9267	8777	4598	3094	3428
4 or more years of c.	10168	10633	5959	3607	3652

Note that the table does not show the totals as given in the text. This is to emphasize that we will not enter totals for the program given below. Commas were also omitted from the numbers because they also will not be entered into the TI-82.

The following program need not be used because percentages can be obtained efficiently, as was demonstrated in the first example in Chapter 1. When you want to make many comparisons, however, the program is more efficient.

The program requires that the data be stored in Matrix [D]. In addition to a list, a matrix is a convenient storage location on the TI-82 for data that has the same number of values in each column, such as the census data table, with four rows of data for each of the five columns.

Entering Data into Matrix (D)

1. Press **MATRX** [EDIT] **4**:[D] and you are in the top line of the matrix editor with the cursor flashing on the number that indicates the number of rows.

2. Type **4** (or if the previous number was two digits press **DEL**) then **ENTER**. The cursor moves over to the number indicating the number of columns.

3. Type **5**, delete any second digit, then press **ENTER** for a screen similar to that on the right. Ignore any values other than zero from previous work because you will be typing over them.

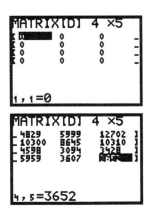

4. Enter the first row by typing 5965 **ENTER** 4755 **ENTER** 4829 **ENTER** 5999 **ENTER** 12702 **ENTER**. The cursor then progresses to the second row. Enter the three other rows of data. Be sure to press **ENTER** after the last entry, 3652, to get the screen on the right. Note that the cursor control keys can be used to move around the editor to check values and make corrections.

Program A6TWTAB

1. Run program A6TWTAB and continue to the first screen on the left at the top of the following page. You will be using only percents, so press **1** for the next screen.

2. At this point we could press **1** to check the inputted values and obtain the column and row totals (see the screen below on the right.)

Be aware of the pause indicator in the upper right corner of the screen throughout this program or you may find yourself waiting for something to happen. The program cannot proceed until you press **ENTER**.

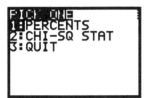 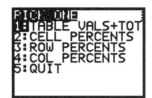 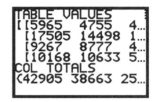

3. If you select **2:CELL PERCENTS** from the option screen you get the screen on the right. The first value of 3.8 indicates that the first value in the last screen, 5965, is 3.8% of all the people in the table.

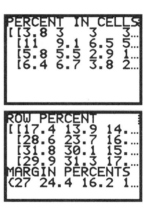

4. If you select **3:ROW PERCENTS** you obtain the screen on the right. The ▶ cursor control key reveals the other values in the rows and shows that the sum of each row equals 100% (e.g., The first row is 17.4 + 13.9 + 14.1 + 17.5 + 37.1 = 100).

5. If you select **4:COL PERCENTS** you obtain the following three screens. Note that the sum of each of the columns adds up to 100%.

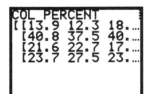 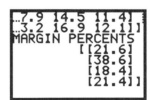

The margin percents (the last screen above) are discussed in Example 2.20 of the text [p. 152], where the first value indicates that in the total survey 21.6% of students did not finish high school. This figure includes all the age groups.

The last row in the COL PERCENT matrix is discussed in Example 2.21 of the text [p. 153], where the first value indicates that 23.7% of those people aged 25 to 34 completed four years of college.

The first and last columns are discussed in Example 2.22 of the text [p. 155]. The first value in the last column indicates that 42.2% of the people over age 65 did not finish high school, and the last value indicates that only 12.1% of them completed four years of college.

We will return to this program in Chapter 8.

CORRELATION COEFFICIENT USING SPREADSHEET

Exercise 2.17 [p. 113]   Find the correlation r step by step, that is, find the mean and standard deviation of the femur lengths given below. Next find the five standardized values for each variable and use the formula for r. Check to see if you get the same value with **STAT**[CALC] 9:LinReg(a+bx)

Femur	(L1)	38	56	59	64	74
Humerus	(L2)	41	63	70	72	84

1. With the femur data in L1 and humerus data in L2, press **STAT**[CALC] 2: 2-Var Stats **L1**, **L2 ENTER** for the screens on the right with $\bar{x}$ = 58.2, Sx = 13.198, $\bar{y}$ = 66 and Sy = 15.890. Note that you need an equal number of values in each list, five in this exercise, to use this calculation.

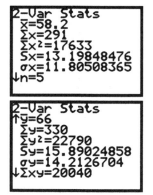

2. Press **STAT 1**:Edit then highlight L3 and type (**L1** - $\bar{x}$) ÷ Sx, as in the first screen below, then **ENTER** for the values in L3 of the second screen. Continue for the next two screens. Note $\bar{x}$, Sx, $\bar{y}$, and Sy can be pasted from **VARS 5:** Statistics...

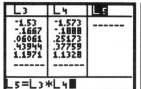

3. Press ↟**QUIT** then ↟**LIST**[MATH] **5**:sum **L5** ÷ 4
ENTER for the screen on the right, with r = 0.994.

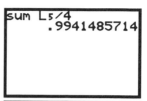

4. Check with **STAT**[CALC] **9**:LinReg(a+bx) **L1**, **L2**
ENTER for the same value of r as above.

5. The two screens below are L1 versus L2 and L3
versus L4. Note the shape is the same but the scale is
quite different.

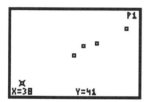 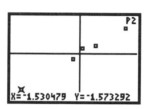

CHAPTER 3
Producing Data

This chapter will show you how to generate random digits and random counting numbers on the TI-82. You will learn how to do this without a program and then learn the appropriate program, A7**RANDOM**, to ease the task.

RANDOM DIGITS AND RANDOM COUNTING NUMBERS

Example [p. 184]   To take a simple random sample of 5 from the 30 businesses listed and numbered on the following page.

To select the sample, work with the random number generator of the TI-82. It is called a pseudo random number generator because it is actually a mathematical recipe or algorithm that is used to calculate these numbers. This recipe needs a number to start with, called a seed. If we all give it the same seed, then we should get the same sequence of values, but there is no detectable pattern that one could use to predict the following values.

We will use a seed in this chapter so you can duplicate the results. There is no need to use a seed in your work unless you are assigned to do so or if you want to duplicate your results. We will use 987 as our seed.

01	A-1 Plumbing	02	Accent Printing
03	Action Sport Shop	04	Anderson Construction
05	Bailey Trucking	06	Balloons Inc.
07	Bennett Hardware	08	Best's Camera Shop
09	Blue Print Specialties	10	Central Tree Service
11	Classic Flowers	12	Computer Answers
13	Darlene's Dolls	14	Fleisch Reality
15	Hernandez Electronics	16	Johnson Commodities
17	JL Records	18	Keiser Construction
19	Liu's Chinese Restaurant	20	Magic Tan
21	Peerless Machine	22	Photo Arts
23	River City Books	24	Riverside Tavern
25	Rustic Boutique	26	Satellite Services
27	Scotch Wash	28	Sewer's Center
29	Tire Specialties	30	Von's Video Store

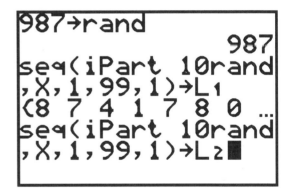

1. From the home screen type 987 then press **STO**▶ followed by **MATH** [PRB] **1**:rand. This is then pasted to the home screen, as in the first line on the screen display. **ENTER** gives 987 as a result, indicating that the seed is set.

2. For the third and fourth lines in the display, engage ▲ **LIST** **5**:seq(, then **MATH** [NUM]**2**: iPart, then type **10**. Press **MATH** [PRB] **1**:rand, then type **,X,1,99,1**). Press **STO**▶ then **L1**. Press **ENTER** and watch the busy signal in the upper right corner of the screen as 99 random digits between 0 and 9 are generated and stored in L1. We see the first few values are 8, 7, 4, 1, 7, 8, 0.

3. To repeat the above for L2, use the **2nd ENTER** feature or ▲**ENTRY**, and the last entry is pasted to the screen. With the ◄ cursor control key, we can enter L2 over L1 as in the last line of the first screen on the following page. Pressing **ENTER** generates 99 more digits but they are stored in L2.

4. To pick our random sample, press **STAT 1:**Edit for the spreadsheet on the right. Use the ▼ cursor control key to go down L1, and we see the sample numbers are 12 (the fourth value down), 09, 26, 07, and 02 ignoring the repeated 12 in the thirteenth row. In order these are business 02, 07, 09, 12, and 26, or Accent Printing, Bennett Hardware, Blue Print Specialties, Computer Answers, and Satellite Services.

Note that it took 15 pairs of digits to come up with our five numbers. This is about what you can expect when you can use only 30% of all possible pairs. The A7**RANDOM** program is more efficient.

L1	L2	L3
8	4	------
7	5	
4	2	
1	2	
7	9	
8	3	
0	9	

L1(1)=8

8	4

L1	L2	L3
5	8	
2	6	
2	2	
5	1	
6	2	
1	7	
0	2	

L1(15)=0

Program A7**RANDOM**

1. Before running this program, set the seed as before with **321➔rand**, so that you can duplicate the results.

2. Run program A7**RANDOM** for the first option screen. Press **1** and the second screen provides the option not to have any repeats.

3. Press **1** and the third screen on the right prompts for **N**, the highest counting number requested.

4. Type **30** then press **ENTER** and the next screen gives 10 and another random value for each press of **ENTER**. After five presses we have, in order, 10, 23, 11, 19, and 13 as shown in the first screen on the following page.

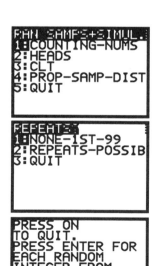

```
RAN SAMPS+SIMUL.
1:COUNTING-NUMS
2:HEADS
3:CLT
4:PROP-SAMP-DIST
5:QUIT
```

```
REPEATS
1:NONE-1ST-99
2:REPEATS-POSSIB
3:QUIT
```

```
PRESS ON
TO QUIT.
PRESS ENTER FOR
EACH RANDOM
INTEGER FROM
1 TO N.
NOTE-NUMS IN L6
N=?30
```

These numbers correspond to the following businesses: Central Tree Service, Classic Flowers, Darlene's Dolls, Liu's Chinese Restaurant, and River City Books.

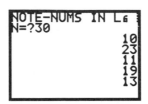

5. Note the pause indicator in the upper right corner of the screen above. To stop the program, press **ON** for the screen below. Press **2** or QUIT. Do not press any other key because you might go to a line in the program listing and inadvertently change the program. If so, you will need to reload it before you can use it again.

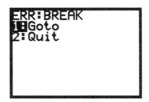

Run the program again. (**ENTER** restarts the program if you have not done anything else on the home screen. If you have, you could use the ↑**ENTRY** feature.) The random numbers 08, 22, 19, 26, and 29 are obtained. Try it! Many random combinations are possible without human bias.

CHAPTER 4

Sampling Distributions and Probability

In this chapter we will simulate taking samples from a population and drawing a distribution of sample proportions and sample means. We will also simulate many tosses of a coin to help explain the ideas of probability. Program A7**RANDOM** will ease this task.

Probability distributions are covered in general and the normal and binomial distributions in particular with the aid of programs A1**PDIST** and A2**BINOM**.

This chapter concludes with an example of using sampling distributions in industry with control charts and program A3**CHART**.

DISTRIBUTION OF SAMPLE PROPORTIONS BY SIMULATION

Simulations of Repeated Samples from a Population with p = 0.6

Example [p. 232]   Simulate drawing 99 samples of 10 and 40 from the population of all U.S. residents, supposing that 60% find shopping time-consuming and frustrating. This work is expanded on in the text where 1000 samples of size 100 and 2500 are taken.

1. Set a seed (as in Chapter 3) 123 **STO▶** rand so you can duplicate the results (see the screen display below right).

2. Generate 10 random numbers between zero and one and store in L1 with seq(rand,X,1,10,1) **STO▶** L1 **ENTER** for {.7039750086 .2..., as on the screen at right.

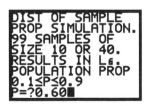

3. Use **↑ENTRY** and the **◄** cursor control key to change L1 to L3, then press **ENTER** for 10 more random values stored in L3.

4. **STAT 1:**Edit reveals the values in the spreadsheet, as on the right. Input a one next to all numbers less than 0.6 (60% in the long run) and a zero next to the rest.

The first sample of ten in L1 has 0100110101, or 5 out of 10 (0.5 or 50%), who feel shopping is time-consuming and frustrating.

The second sample of ten in L3 has 1011110110 or 7 out of 10 (0.7 or 70%).

5. To speed up this sampling with a program, first reset the seed with 123 **STO▶** rand.

6. Run program A7**RANDOM 4:**PROP-SAMP-DIST to get the display screen on the right, which has the P=? prompt in the last line.

7. Type **0.60 ENTER** for the next screen, below.

8. Type **1** for samples of 10, and the screen on the right fills up with these samples. Note that the first two rows are the same as the two samples in step 4 above and that we have one 0.4, two 0.5, three 0.6, and two 0.7 values.

9. Press **ENTER** to bring up the histogram of the eight samples above, as in the screen on the right. Although it is small, the trace feature indicates the three values between 0.55 and 0.65.

10. Press **ENTER** again to bring up the numbers 9, 10, ... 98, 99 to the screen as these 99 samples are being calculated and the proportion who feel shopping is time-consuming and frustrating are stored in L6. (This may seem slow, but it beats doing it by other ways.) The summary screen is then given. The mean of the 99 samples was 0.608 compared to the population value of 0.6 and the standard deviation 0.14 compared to 0.155.

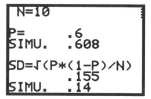

11. Press **ENTER** now to bring up the histogram of all 99 samples and $(27+27+20)/99 = 72.7\%$ of the values are between 0.45 and 0.75.

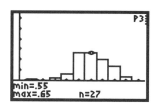

12. Press **ENTER** to fit a normal curve using the population values given a few screens back. (See the screen below.)

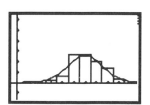

13. Press **ENTER** to run the simulation with samples of 40 (take a break while waiting), and bring up the screen on the right, which has about the same mean as before but with about half the spread.

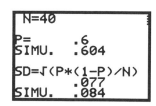

14. Press **ENTER** to reveal that there is now (18+46+26)/99 = 90.9% of the values between 0.45 and 0.75, whereas before there was about 72.7%. Compare the spread between the fifth screen on the previous page and the two screens on the right.

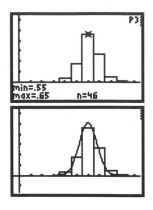

As you can see, using small samples to try to predict the population proportion can distort the population parameter by more than 0.15 about 10% of the time for a sample of size 40. We now see why the text uses sample sizes of 2500 in constructing its distributions: because it has much less spread and thus is a more accurate predictor.

Note: A solution to Exercise 4.7 of the text [p. 240] is given at the end of this chapter.

RANDOMNESS AND PROBABILITY

"Random" in statistics does not mean haphazard but rather refers to a kind of order that emerges only in the long run.

Example [p. 242]   We can duplicate on the TI-82 the text example of tossing coins, where the proportion of heads is quite variable for only a few coin tosses but approaches 0.5 after many tosses.

1. Set the seed with 4567 **STO**▶rand

2. Generated 99 ones and zeros in L1 with seq(iPart 2rand,X,1,99,1 **STO**▶L1 **ENTER** (similar to what was done in Chapter 3).

3. The screen on the right shows that the first value is a one (or the first toss a head) for 100% heads. This is follwed by another toss of a head for 100% heads. After five tosses there are 3 heads and 2 tails for 60% heads. Using the ▶ cursor key, you see seven of the first ten values are ones or 70% heads, still quite variable.

4. You could continue with the above but see what happens after 99 tosses with sum L1/99 (sum under ↑ **LIST** [MATH]) for 0.505, as in the last line of the previous screen display.

5. Reset the seed, 4567 **STO▶**rand, and run program A5**RANDOM** 2:HEADS for the instruction display screen below on the left.

6. Press **ENTER** to get the first line at the left with a head, H, on the first toss for 100% heads. Continue to press **ENTER** until after 5 tosses there are 60% heads, just as in step 3 and in the screen below on the right.

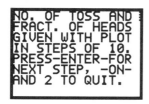   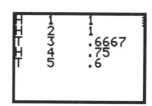

7. After ten tosses there are 70% heads and the results are plotted as in the screen on the right.

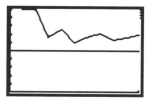

8. Press **ENTER** to continue with 10 more tosses with only the toss number and proportion of heads shown. After 20 tosses there are 55% heads in the screen below on the left.

9. Press **ENTER** to bring up the plot of the first 20 points.

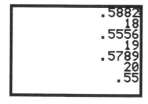   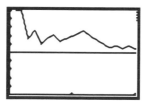

10. The program continues, plotting after every ten tosses until after 90 tosses the plot obtained has 50% heads (for this particular seed). Note that the proportion of heads was less than 50% at around 30 tosses.

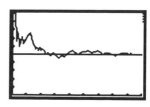

11. Press **ENTER** to continue the program without any plots, until after 1000 tosses there are 502 heads. Try it again for another variation of the results.

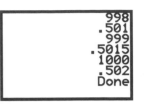

MEAN AND STANDARD DEVIATION OF A DISCRETE RANDOM VARIABLE

Example [p. 258]   Find the mean and standard deviation of the following distribution:

Household size (L1)	1	2	3	4	5	6	7
Probability (L2)	0.251	0.321	0.171	0.154	0.067	0.022	0.014

1. Enter the above values into a spreadsheet. Highlight L3 by moving the cursor there (as in the first screen below) and enter **L1*L2**. **ENTER** gives the product of household size by the probability of obtaining that household, as shown in the second screen below.

2. Highlight L4 and subtract the mean household size (sum L3) from each household size and square it, as in the two screens on the right below.

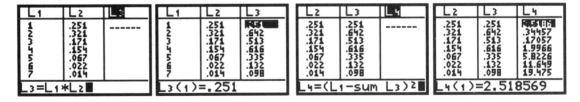

3. Multiply the differences squared in L4 by the probabilities in L2 for the two screens below.

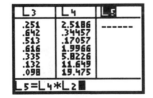

4. Press ↟**QUIT** to return to the home screen, where you sum L2 (sum under ↟**LIST** [MATH]) to show that it adds to one and is truly a probability distribution. Next, sum L3 for the mean of this distribution, and sum L5 for the variance. All of these are shown in the screen on the right and in the text [pp. 258, 260].

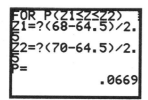

NORMAL DISTRIBUTION

Example [p. 263]   If we look at the heights of all young women, we find that they closely follow the normal distribution with mean = 64.5 inches and a standard deviation = 2.5 inches. What is the probability that a randomly chosen young woman is between 68 and 70 inches tall?

This reduces to finding the area under a normal curve as in Chapter 1 with program A1**PDIST** 1:Z-TO-P, with the input and output as shown in the screens below.

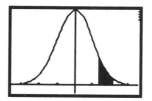

SAMPLE PROPORTIONS

Standard Deviation for the Sampling Distribution of p

Example [p. 270]   For p = 0.35 this is calculated on the TI-82 with $\sqrt{}$ (p(1–p)/n) or as $\sqrt{}$ (0.35*0.65/1500) for 0.0123, as in the screen on the right and in the text.

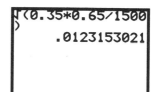

THE BINOMIAL DISTRIBUTIONS

Example (p. 283)   Each child born to a particular set of parents has probability 0.25 of having blood type O. If these parents have five children, what is the probability that exactly two of them have type O blood?

1. 10*0.25^2*0.75^3 **ENTER** for 0.2637, as in the screen on the right; or

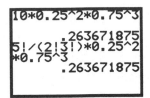

2. 5!/(2!3!)*0.25^2*0.75^2 **ENTER** for 0.2637 as in the screen on the right (with ! under **MATH**[PRB]4:!); or

3. **5** nCr **2** *0.25^2*0.75^3 **ENTER** for 0.2637, as in the screen at right, with nCr under **MATH**[PRB]3:nCr; or

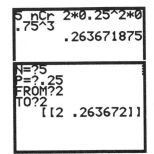

4. Program A2**BINOM**.
Run the program and input 5 for N, 0.25 for P and go FROM 2 TO 2 (this will be clearer in the next example) for the result [[2 .263672]], which indicates that the probability of exactly 2 successes is 0.2637.

Example [p. 285]   The probability that no more than one switch fails out of 10 switches, with a probability of failure of p = 0.10 is P(X = 0) + P(X = 1).

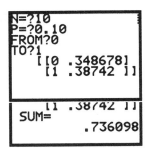

1. Run the program **A2BINOM** and input values for N and P and indicate you want FROM 0 TO 1 failed switches for the first screen on the right. **ENTER** again brings the partially hidden screen on the right, with the sum of the two probabilities being 0.736098, or P(X = 0) = 0.348678 and P(X = 1) = 0.38742, and the sum of these two is 0.7361.

2. Press **ENTER** then choose the Plot option for the plot shown below and as in the text [p. 288]. Note the cursor key was used to highlight probability of P(X = 1) of approximately 0.39 (to the closest percent).

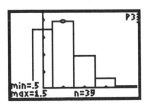

3. Press **ENTER** and a normal curve is fitted to the data (shown below) and followed by a shaded area that we might hope to approximate the binomial probabilities under the right conditions. These conditions are clearly not met here, in that the binomial distribution is not even symmetrical.

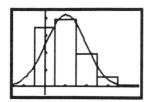
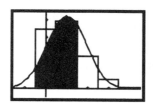

The following does much better with n = 100 and p = 0.10, so that the conditions (rule of thumb) of the text [p. 272] are met, with np = 100*0.10 = 10 and n(1 − p) = 100*0.9 = 90 both greater than or equal to 10. Note that this graph is approximately symmetrical about its mean (10), but many cells are not shown because their probabilities are so small. For example, the probability of getting 20 or more failed switches out of 100 switches is very unlikely when there are only 10% defective switches in the population. As np and n(1 − p) become larger, the normal approximation becomes better.

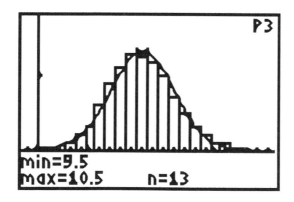

SAMPLE MEANS AND THE CENTRAL LIMIT THEOREM

To investigate the distribution of sample means, we will toss nine-sided dice. The wonders of simulations!

1. On the home screen first set the seed so you can duplicate the results, if you wish, by keying (with rand under **MATH** [PBR]) 4321 **STO▶** rand **ENTER**.

2. Using seq (under ↑ **LIST**) and iPart (under **MATH** [NUM]), type, paste, and enter the following (do not forget the last entry feature of ↑ **ENTRY**):

seq(iPart 9rand+1,X,1,99,1 **STO▶ L1 ENTER**
seq(iPart 9rand+1,X,1,99,1 **STO▶ L2 ENTER**
seq(iPart 9rand+1,X,1,99,1 **STO▶ L3 ENTER**
seq(iPart 9rand+1,X,1,99,1 **STO▶ L4 ENTER**
seq(iPart 9rand+1,X,1,99,1 **STO▶ L5 ENTER**

3. Next, find the mean of each position of these lists by the following: (**L1+L2+L3+L4+L5**)/5 **STO▶ L6 ENTER**.

4. With **STAT 1**:Edit... use the ▼ cursor control key to go down to the sixtieth row to show the following screens.

Note that the first five values in the sixtieth row are large and their mean = (9 + 8 + 7 + 9 + 9)/5 = 8.4, while the first five values in the fifty-fourth row (the top row shown on the screens) are small (2,3,3,5,2) with an average of 3. It would be very unlikely to get all 1's or all 9's. The fifty-fifth row is much more typical with 1,5,6,8,9 for a mean of 5.8 and with the small value of 1 balanced by a large value of 8 or 9. Note that we went to the sixtieth row to show all the above with the minimum amount of output.

5. To summarize the data, plot some histograms with the window and setup (at the top of the following page on the left). First, plot the histogram of L1, which should be fairly uniform (at the top of the following page on the right) with about 11 of each digit 1,2,3,4,5,6,7,8,9 because 99 digits were generated for each list. Set Y1=11 after pressing the **Y=** key and then **TRACE**. Histograms of L2, L3, L4, and L5 would look similar.

WINDOW FORMAT	Plot1	Y1■11	P1
Xmin=.5	▓ Off	Y2=	
Xmax=9.5	Type:⌐ ∠ ⏣ ⣿	Y3=	
Xscl=1	Xlist:█L1 L2 L3 L4 L5 L6	Y4=	
Ymin=-4.5	Freq:█L1 L2 L3 L4 L5 L6	Y5=	
Ymax=45		Y6=	
Yscl=0		Y7=	
		Y8=	

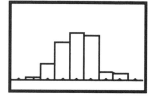

6. By setting up for L6 (and turning off or clearing Y1), the histogram on the right appears. Note that this is much more normally shaped with no means of 1 (the first cell) or 9 (the last cell) as conjectured. Most of the means are lumped near the center at 4, 5, and 6.

All of these procedures have been automated in the program **A7RANDOM** 3:CLT, CLT standing for Central Limit Theorem, which gives the option of tossing one, two, four, and eight dice 99 times.

Program A7RANDDOM 3: CLT (Central Limit Theorem Simulation)

1. First, set the seed 4321 **STO▶** rand **ENTER**.

2. Run the program, and after reading the intro screen, press **4** when the following screen below left appears to demonstrate tossing a nine-sided die 8 times (or 8 die once) and find the mean of the 8 values, which are shown in the middle screen as $(6+4+8+5+2+9+9+3)/8 = 5.75$. This is repeated for a total of eight means, then press **ENTER** for the histogram of these eight means, shown below right, with one value near 2 (2.125 between 1.5 and 2.49), two values near 4 (3.875, 4.25), one value near 5 (5.375) and the other four values near 6.

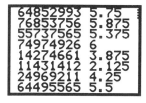

3. Press **ENTER** again and the screen prints out 9, 10, 11,...97, 98, 99, while each of the other 91 simulations are being carried out. The screen on the right gives the mean of all 99 means (5.035 in this example), compared with the theoretical value of 5 and the standard deviation for all 99 means (0.88 in this example), compared to the theoretical value of 0.913.

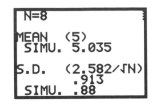

4. Press **ENTER** again and the histogram on the right is plotted (with a normal curve of the same mean and standard deviation). More of the means are lumped near the center values of 4, 5, and 6 as predicted by the Central Limit Theorem.

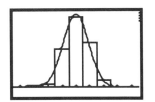

The result for 5 dice is repeated on the right with the same window to ease the comparison. With the larger sample size, the spread of the means is reduced. Try the program with different numbers of dice to convince yourself of the feasability of the Central Limit Theorem.

CONTROL CHARTS

Example [pp. 306–308]   Twenty means of sample of 4 measuring the tension of a computer monitor component is repeated in the table on the right. Also given is the desired centerline, the desired tension of 275 mV and the lower (210.5) and upper (339.5) control limits.

Y1=339.5	Plot1	WINDOW FORMAT
Y2=275	▓▓ Off	Xmin=0
Y3=210.5	Type: ▰ ⌐ ▱ ▥	Xmax=22
Y4=	Xlist: **L1** L2 L3 L4 L5 L6	Xscl=5
Y5=	Ylist: L1 **L2** L3 L4 L5 L6	Ymin=150
Y6=	Mark: ▢ ＋ ・	Ymax=400
Y7=		Yscl=50
Y8=		

1. Enter the data in L1 and L2.

Samp(L1)	mean(L2)
1	269.5
2	297
3	269.5
4	283.3
5	304.8
6	280.4
7	233.5
8	257.4
9	317.5
10	327.4
11	264.7
12	307.7
13	310
14	343.3
15	328.1
16	342.6
17	338.8
18	340.1
19	374.6
20	336.1

2. Put the centerline and control limits in Y1, Y2, Y3, under **Y=** with other plots off as shown in the left screen on the previous page.

3. Define plot 1 with **↑STAT/PLOT**, as in the center screen on the previous page.

4. Set **WINDOW** as in the right screen on the previous page.

5. Press **TRACE** and ► a few times for the control chart shown below, with the 14th value going out of control.

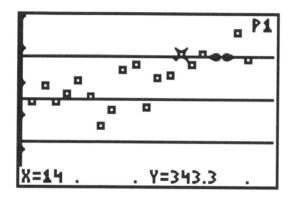

Program A3**CHART**

The preceding procedure is automated with program A3**CHART**, which gives the first screen below to remind us to put the data in L1 and L2. Next, input the appropriate values, as in the middle screen. Press **ENTER** to get the control chart. Press **TRACE** and the ► key for the screen on the right.

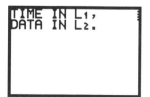

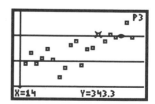

Program UTILS for Data Sets Larger Than 99 Values

Exercise 4.7 [p. 240]   The following data set has 100 values and causes a problem because a list can hold only 99 values. Find $\bar{x}$, Sx and a histogram.

17	23	18	27	15	17	18	13	16	18	20	15	18	16	21
17	18	19	16	23	20	18	18	17	19	13	27	22	23	26
17	13	16	14	24	22	16	21	24	21	30	24	17	14	16
16	17	24	21	16	17	23	18	23	22	24	23	23	20	19
20	18	20	25	16	24	24	24	15	22	22	16	28	15	22
9	19	16	19	19	25	24	20	15	21	25	24	19	19	20
28	18	17	17	25	17	17	18	19	18					

If we transfer the date file ex4-7.dat from a computer to a TI-82 (covered in Appendix A), the first 99 values read one row at a time could be loaded to a list, which would only leave off the last value of 18. The data in L1 in the screen on the right shows the 93rd to 99th values from the last row of above data. (This data could also be entered by hand.)

When we construct the histogram from L1 shown below right, with the window setting below left, only one value is missing, an 18. Thus, the cell that has been traced should have 12 instead of the 11 values shown. Note that dividing each value by 200 does not change the shape of the histogram.

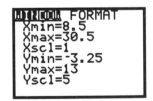

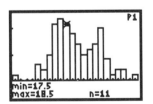

1-Var Stats of L1 gives Σx = 1944, so that the mean of all 100 values is $(1944 + 18)/100 = 19.62$ and Sx for all

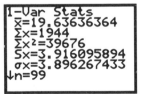

100 values could be calculated as $\sqrt{((39676 + 18^2 - (1944 + 18)^2/100)/99)} = 3.899702138$. Note that $19.62/200 = 0.0981 = 0.10$.

A more general method (one that works for larger data sets) is to transfer the data file to a matrix, as shown on the right. Note that this matrix has 50 rows and 2 columns, with the first row being the first two values in the first row of our original data, the 2nd row the next two values, and so on. Instead of a 50 x 2 matrix, it could have been 25 x 4, 20 x 5, or 10 x 10, but because these columns will be transferred to lists, you want the fewest number of columns.

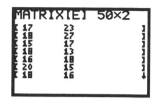

If your data was originally transferred to Matrix [E], store it in Matrix [D] with [E] **STO►** [D] because it is needed for the next step.

Program AA0UTILS

1. To calculate summary statistics

Store this matrix into [D] and run program AA0UTILS for the option screen on the right. Press **3** and, after some calculation time, the following screen of output appears with the mean and standard deviation of all 100 values.

2. To transfer column of matrix to L6 for histograms

With the data in Matrix [D] run program AA0UTILS 2:COL-[D]-TO-L6 to get the screen on the right. Type **1** at the prompt and the first column of the data Matrix [D] is stored in L6.

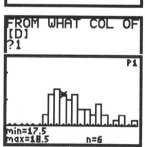

Construct a histogram of the data in L6 and with the **TRACE** key make a frequency table for each value. Note that there are 6 eighteens.

Repeat the step for the second column of [D] to get the histogram on the right with 6 more eighteens.

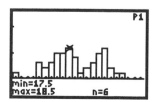

Combine the above two frequency tables to get the histogram of all 100 values as in the screen on the right.

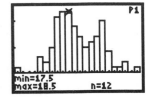

Note that the program also allows the transfer from a list to a column of a matrix.

CHAPTER 5
Introduction to Inference

In this introduction to inference a one sample mean is used to estimate the population mean with a confidence interval, or the sample mean is used to test a hypothesis about the population mean. The calculations required are straightforward and shall be done on the home screen. Program A1**PDIST** is used to find the values of the tail probabilities of the normal distribution and help in making decisions.

The examples in this chapter are from the text, which you should study for the full details of the problems and the requirements that must be met for the calculations to have the intended meaning.

CONFIDENCE INTERVALS

Example [p. 325]   A simple random sample of 840 young men yields a mean score of $\overline{x} = 272$ on a NAEP test of quantitative skills. If it is known from past experience that the standard deviation of scores in the population of all young men is $\sigma = 60$, find the 95% confidence interval for μ.

Estimate ± margin of error

$\bar{x} \pm z^*\sigma/\sqrt{n}$

$272 \pm 2(60/\sqrt{840})$

272 ± 4.14

```
2*60/√840→M
          4.140393356
272-M
          267.8596066
272+M
          276.1403934
```

or 267.9 to 276.1, as shown in the screen on the right, and close to the approximation in the text [p. 328].

Note that you can use ↑ENTRY and change − to + for the last calculation.

Level C Confidence Intervals and Critical Values

Example [p. 332]   Find z* for an 80% confidence interval.

1. $(1 - C)/2 = (1 - 0.80)/2 = 0.20/2 = 0.10$ in each tail.

2. Run program A1**PDIST** 2:P-TO-Z and enter .10 at the prompt for a z value of ⁻1.28. By repeating this for P = .90, you obtain 1.28, as expected from symmetry, so z* = 1.28. The screen outputs below show where the plot options were also used.

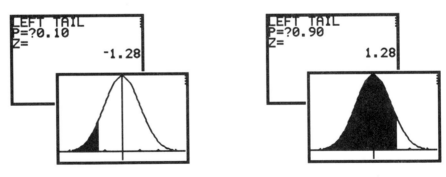

A table of common confidence levels are given in the text [p. 334] and are duplicated at the top of the following page with an extra column for the output of program A1**PTAILS** 2:P-TO-Z, which only gives output to two decimal places and the second decimal is not always accurate. The output of program A1**PTAILS** 3:Z-DIST, used later in this chapter, is accurate to the four places given.

Confidence level %	Tail area	z^*	PrgmP-TO-Z
90	0.05	1.645	1.64
95	0.025	1.96	1.96
99	0.005	2.576	2.58

Example [p. 335]   The results of repeated measurements of concentration of the active ingredient of a pharmaceutical product follow a normal distribution quite closely, with $\sigma = 0.0068$ grams per liter. If three analyses of one specimen gives concentrations of 0.8403, 0.8363, and 0.8447:

1. Find the 99% confidence interval for the true concentration of μ.

The mean of three sample values is approximately 0.8404 and the critical value for C = 0.99 is 2.576. Thus,

```
2.576*0.0068/√3→
M
           .0101133292
0.8404-M
           .8302866708
0.8404+M
           .8505133292
```

$x \pm z^*\sigma/\sqrt{n} = 0.8404 \pm 2.576{*}0.0068/\sqrt{3}$

0.8404 ± 0.0101

(0.8303, 0.8505)

2. [p. 339] Find the 90% confidence interval for the true concentration of μ. Press the last ↑ENTRY feature a few times to call up the margin of error calculation and type 1.645 over the 2.576. Then press ↑ENTRY again to recall the $0.8404 \pm M$ to get the screen on the right with the smaller interval below (for a minimum amount of typing).

```
1.645*0.0068/√3→
M
           .0064582401
0.8404-M
           .8339417599
0.8404+M
           .8468582401
```

(0.8339,0.8469)

CHOOSING THE SAMPLE SIZE

Example [p. 342]   Management wants a margin of error m = 0.005 with 95% confidence. How many measurements must be averaged to comply with this request?

```
(1.96*.0068/.005
)²
           7.10542336
```

$n = (z^*\sigma/m)^2 = (1.96 * 0.0068/0.005)^2 = 7.1$, thus 8

TESTS OF SIGNIFICANCE AND P-VALUES

Example [p. 349]   Diet colas use artificial sweeteners to avoid sugar. Did the cola under study lose sweetness under storage? Sweetness losses (and two gains) measured by ten trained testers are:

2.0 0.4 0.7 2.0 ⁻0.4 2.2 ⁻1.3 1.2 1.1 2.3

From experience it is known that the scores of individual tasters vary according to a normal distribution with $\sigma = 1$.

Test:

H_0: $\mu = 0$ (no change of mean sweetness)

H_a: $\mu > 0$ (a loss of mean sweetness or a significant positive difference)

1. Store the 10 differences in L1 and use 1-VarStats L1 for $\overline{x} = 1.02$.

2. $\sigma/\sqrt{n} = 1/\sqrt{10} = 0.316$, one standard deviation of the normal distribution of sample means.

3. The z value for our sample means is $(1.02-0)/0.316 = 3.23$.

4. Running program A1**PTAILS** 3:Z-DIST reveals a right-tail p-value of 6E ⁻4 = 6*10⁻⁴ = 0.0006 (see the screen on the right).

Ten tasters would have an average score as large as 1.02 only 6 times in 10,000 tries if the true sweetness change were 0. An outcome this unlikely convinces us that the true mean is really greater than 0 [pp. 353, 354].

Example [p. 360]   If the ten tasters had an $\overline{x} = 0.3$, this would have a z value of $z = (0.3 - 0)/0.316 = 0.95$ with a p-value of about 0.17, as in the screen on the right. An outcome this likely to occur just by chance is not good evidence against the null hypothesis [p. 353].

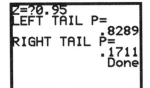

Two-Sided Test (p. 357)

Example [p. 366]   The National Center for Health Statistics reports the mean systolic blood pressure for males 35 to 44 years of age is $\mu = 128$, and the standard deviation of the population is $\sigma = 15$. A large company finds the mean systolic blood pressure of a sample of 72 executives in this age group is $\bar{x} = 126.07$. Is this evidence that the company's executives have a different mean blood pressure from the general population?

1. $H_o: \mu = 128$ $H_a: \mu \ne 128$

2. $z = (\bar{x} - \mu_0)/(\sigma/\sqrt{n}) = (126.07 - 128)/(15/\sqrt{72}) = {}^{-}1.09$

3. Running program A1**PTAILS 3**:Z-DIST reveals a left-tail probability of 0.1379. Thus, the P-value = 2*0.1379 = 0.2758.

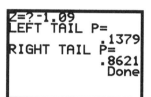

Conclusion: More than 27% of the time, an SRS of size 72 from the general male population would have a mean blood pressure at least as far from 128 as that of the executive sample. The observed $\bar{x} = 126.07$ is therefore not good evidence that executives differ from other men [pp. 366, 367].

Test with Fixed Significance Level (p. 370)

Example [p. 371]   Refer to the pharmaceutical product example earlier in this chapter. The analytical laboratory is asked to evaluate the claim that the concentration of the active ingredient in a specimen is 0.86%. The lab makes 3 repeated analyses of the specimen. The mean result was $\bar{x} = 0.8404$. Given is $\sigma = 0.0068$. Is there significant evidence at the 1% level that $\mu \ne 0.86$?

1. $H_o: \mu = 0.86$ $H_a \mu \ne 0.86$ $\alpha = 0.01$

2. $z = (0.8404 - 0.86)/(0.0068/\sqrt{3}) = {}^{-}4.99$

3. $z^* = 2.576$ from Table C of the text.

$4.99 > 2.576$ or $^-4.99 < ^-2.576$.

Conclusion: We reject the null hypothesis and conclude (at the 1% significance level) that the concentration is not as claimed. It is statistically significantly less.

Note that the left-tail probability is given as 0.0000^+ (from the output screen of A1**PTAILS** 3:Z-DIST on the right) or a value less than 0.00005. Since this is a two-tail test, when we multiply by 2 we obtain a value less than 0.0001.

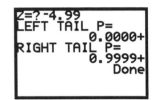

Two-Sided Test from Confidence Intervals (p. 374)

Example [p. 375]   The 99% confidence interval for μ is $x \pm z^*\sigma/\sqrt{n} = 0.8404 \pm 2.576*0.0068/\sqrt{3} = 0.8404 \pm 0.0101$ or $(0.8303, 0.8505)$.

The hypothesized value $\mu_0 = 0.86$ above falls outside this confidence interval, so we reject H_0: $\mu = 0.86$ at the 1% significance level.

On the other hand, we cannot reject H_0: $\mu = 0.85$ at the 1% level in favor of the two-sided alternative H_a: $\mu \neq 0.85$, because 0.85 lies inside the 99% confidence interval for μ. $(0.8303 < 0.8500 < 0.8505) < 0.8600$.

CHAPTER 6
Inference for Distributions

Inference from one sample mean in this chapter will be similar to Chapter 5, but here we use the program A1**PDIST** 4:T-DIST. Inference from two sample means will be aided by program A5**TSAMPS** and from two sample standard deviations by program A1**PDIST** 6:F-DIST.

Study the text to learn the proper procedures for checking the assumptions for the test used in this chapter.

INFERENCE FOR THE MEAN OF A POPULATION

The t Confidence Intervals (p. 411)

Example [p. 413]   Five cockroaches fed the sugar D-glucose and dissected after 10 hours had the following amounts (in micrograms) of D-glucose in their hindguts:

55.95 68.24 52.73 21.50 23.78

Find the 95% confidence interval for the mean amount of D-glucose in cockroach hindguts under these conditions.

1. With the data in L1 use 1-VarStats L1 for $\overline{x}$ = 44.44 and s = 20.741 as in the first screen on the right.

```
1-Var Stats
x̄=44.44
Σx=222.2
Σx²=11595.2914
Sx=20.74080158
σx=18.55113689
↓n=5
```

2. With t* = 2.776 from Table C of the text for D.F. = 5 – 1 = 4 calculate the margin of error and store as M.

$t*(s/\sqrt{n}) = 2.776 \times 20.741 \div \sqrt{5}$ **STO▶** M

3. The 95% CI = x ± M = 44.44 ± M = 44.44 ± 25.75 = (18.69, 70.19), as in the screen on the right.

```
2.776*20.741/√5→
M
           25.74922434
44.44-M
           18.69077566
44.44+M
           70.18922434
```

Hypothesis t Test (p. 412)

Example [p. 414]   Cola makers test new recipes for loss of sweetness during storage. The sweetness losses (sweetness before storage minus sweetness after storage found by 10 tasters for one new cola recipe are as follows:

2.0 0.4 0.7 2.0 ⁻0.4 2.2 ⁻1.3 1.2 1.1 2.3

Are these data good evidence that the cola lost sweetness?

1. $H_o: \mu = 0$ $H_a: \mu > 0$.

2. With the data in L1 use 1-VarStats L1 for $\overline{x}$ = 1.02 and s = 1.196 as in the screen at the right.
$t = (x - \mu_0)/(s/\sqrt{n}) = (1.02 - 0)/(1.196/\sqrt{10}) = 2.70$ as in the screen below.

```
1-Var Stats
x̄=1.02
Σx=10.2
Σx²=23.28
Sx=1.196104789
σx=1.134724636
↓n=10
```

```
(1.02-0)/(1.196/
√10)
           2.696925764
```

3. Run program A1**PDIST 4:**T-DIST and input 2.7 for T and $10 - 1 = 9$ for D.F. for a right-tail probability or p-value = 0.0122, as shown on the right.

```
T=?2.7
D.F.=?9
LEFT TAIL P=
           .9878
RIGHT TAIL P=
           .0122
           Done
```

Conclusion: There is quite strong evidence for a loss of sweetness [p. 414].

Matched Pairs t Procedure

Example [p. 419]   Twenty teachers at a summer institute were give a test before and after a four-week course in spoken French. Assess whether the institute significantly improved the teachers' comprehension of spoken French.

1. Test H_o: $\mu = 0$ H_a: $\mu > 0$.

2. Do the t test on the gains (or losses) in test scores from the pretest to the posttest.
(a) With the data in L1 and L2 and with L3 highlighted, as in the first screen below, type **L2 − L1** then **ENTER** for the gains in L3, as shown in the screen partly hidden behind the first.
(b) Use 1–VarStats L3 for the mean and standard deviation of the gains $\bar{x} = 2.5$ and s = 2.893 as in the middle screen.
(c) $t = (\bar{x} - 0)/(s/\sqrt{n}) = (2.5 - 0)/(2.893/\sqrt{20}) = 3.86$.

	L1	L2
Teacher	Pretest	Posttest
1	32	34
2	31	31
3	29	35
4	10	16
5	30	33
6	33	36
7	22	24
8	25	28
9	32	26
10	20	26
11	30	36
12	20	26
13	24	27
14	24	24
15	31	32
16	30	31
17	15	15
18	32	34
19	23	26
20	23	26

```
L1    L2    L3       L3
32    34    ------    
31    31             0
29    35             6
10    16             6
30    33             3
33    36             3
22    24             2
L3=L2-L1
```

```
1-Var Stats
x̄=2.5
Σx=50
Σx²=284
Sx=2.892822333
σx=2.819574436
↓n=20
```

```
(2.5-0)/(2.893/√
20)
        3.864618005
```

3. Run program A1**PDIST** 4:T-DIST and input 3.86 for T and $20 - 1 = 19$ for D.F. for a right-tail probability or p-value = 5E ⁻4 = 0.0005, as shown in the screen display on the right.

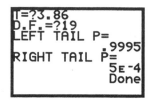

Conclusion: The improvement in listening scores is very unlikely to be due to chance alone. We have strong evidence that the institute was effective in raising scores [pp. 420, 421].

Note: Just as the above procedure was similar to the previous example, you could calculate the confidence intervals for the gains similar to the procedure described in the first example of this chapter.

COMPARING TWO MEANS

Example [p. 439]   Does increasing the amount of calcium in our diet reduce blood pressure? An experiment was run with two groups. One group received calcium supplements and the other a placebo. The response variable is the decrease in systolic blood pressure in millimeters of mercury for a subject after 12 weeks. An increase appears as a negative response.

L1 Calcium Group   7 ⁻4 18 17 ⁻3 ⁻5 1 10 11 ⁻2
L2 Placebo Group   ⁻1 12 ⁻1 ⁻3 3 ⁻5 5 2 ⁻11 ⁻1 ⁻3

Hypothesis Test (p. 443)

1. With the data in L1 and L2, use 1-VarStats for the mean and standard deviations summarized in the following table.

Group	Treatment	n	mean	s
1	Calcium	10	5	8.743
2	Placebo	11	-0.273	5.901

The calcium group shows a mean drop in blood pressure of 5.0, while the placebo group had almost no change, ⁻0.273. Is this outcome good evidence that calcium decreases blood pressure in the entire population of healthy black men more than the placebo does?

2. The test statistic for the null hypothesis $H_0: \mu_1 = \mu_2$ is obtained most easily by running program A5**TSAMPS** 1:MEANS(T-TEST), which needs the summary statistics from the previous table for input, as in the screen on the right.

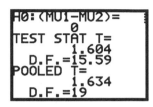

```
SAMPLE 1
MEAN=?5
S.D.=?8.743
N1=?10
SAMPLE 2
MEAN=? -0.273
S.D.=?5.901
N2=?11■
```

3. Pressing **ENTER** after entering 11 for N2 gives the second screen, with the test statistic t = 1.604. (Note that the text discusses the pooled t on p. 455 but it is not covered here.)

```
H0:(MU1-MU2)=
            0
TEST STAT T=
        1.604
    D.F.=15.59
POOLED T=
        1.634
    D.F.=19
```

4. To find the p-value, the text uses a conservative value for D.F. = 10 − 1 = 9 (the smaller sample size minus one) [p. 444], but discusses the D.F. = 15.59 of the last screen [p. 451].

Run program A1**PDIST** 4:T-DIST for the first case (D.F. = 9), which gives a one-tail p-value = 0.0716, and the second a p-value = 0.0648, as in the screens on the right.

```
T=?1.604
D.F.=?9
LEFT TAIL P=
        .9284
RIGHT TAIL P=
        .0716
```

```
T=?1.604
D.F.=?15
LEFT TAIL P=
        .9352
RIGHT TAIL P=
        .0648
        Done
```

Note: D.F. = 15.59 would be treated the same way as 15 in the program (it takes the integer part), giving us a conservative p-value as mentioned in the text [p. 442].

Conclusion: The experiment found evidence that calcium reduces blood pressure, but the evidence falls a bit short of the traditional 5% and 1% levels [p. 444].

Note: The t-value could have been obtained on the home screen as in the screen below left, and the degrees of freedom as in the screen below right.

```
8.743²/10→A
        7.6440049
5.901²/11→B
        3.165618273
(5- -0.273)/√(A+B
)
        1.603808438
```

```
(A+B)²/(A²/9+B²/
10)
        15.59131407
```

Confidence Interval for μ_1 - μ_2 (p. 442)

Continuing with the above example, find the 90% confidence interval of the mean advantage of calcium over a placebo, μ_1 - μ_2 [pp. 444 and 451].

Continuing with program A5TSAMPS 1:MEANS(T-TEST) after doing the hypothesis test above, the choices in the first screen on the right appear.

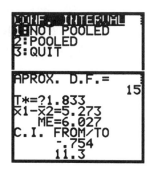

Select 1 and (a) entering the critical t* = 1.833, that goes with the conservative D.F. = 9, we obtain the interval (⁻0.754, 11.3), as shown in the second screen on the right and in the text [p. 444].

(b) Entering the critical t* = 1.753 for D.F. = 15 gives the interval (⁻0.491, 11.04), as in the screen on the right and in the text [p. 451].

Note: Continue with the home screen example for the screen below.

```
1.753√(A+B)→M
       5.763511889
(5--0.273)-M
      -.4905118892
(5--0.273)+M
       11.03651189
```

THE F TEST FOR COMPARING TWO STANDARD DEVIATIONS

Continue with the above example to compare standard deviations to see whether calcium changes the variation in blood pressures [p. 466].

1. Test H_o: $\sigma_1 = \sigma_2$ Ha: $\sigma_1 \neq \sigma_2$.

2. The larger of the two sample standard deviations is s = 8.743 from ten observations. The other is s = 5.901 from eleven observations (as in the

previous table). The F test statistic is therefore

$$F = \text{larger } s^2/\text{smaller } s^2 = 8.743^2/5.901^2 = 2.195$$

3. To find the p-value, run program A1**PDIST** 6:F-DIST and enter 2.195 for F, $10 - 1 = 9$ for the numerator df, and $11 - 1 = 10$ for the denominator D.F. for the right-tailed probability of 0.118, as in the screen on the right, which gives a p-value = 2*0.118 = 0.236.

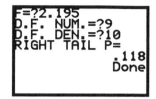

Conclusion: If the populations were normal, the observed standard deviations would give little reason to suspect unequal population standard deviations. Because one of the populations shows some nonnormality, we cannot be fully confident of this conclusion [p. 466].

CHAPTER 7
Inference for Proportions

This chapter will do inference on proportions from one and two independent samples. For the two-sample case we will be aided by program A5**TSAMPS.**

INFERENCE FOR ONE POPULATION PROPORTION

Confidence Interval (p. 490)

Example [p. 484]   Behavioral surveys interviewed a random sample of 2673 adult heterosexuals. Of these 170 had more than one sexual partner in the past year or $\hat{p} = 170/2673 = 0.0636$ or 6.36% (stored as P on the screen on the right).

```
170/2673→P
           .0635989525
2.576√(P*(1-P)/2
673)→M
           .0121591245
P-M
           .051439828

P+M
           .075758077
```

Find the 99% confidence interval for the proportion p of all adult heterosexuals with multiple partners based on the above sample. The critical value $z^* = 2.576$ is from Table C in the text. We want $\hat{p} \pm z^* \sqrt{(\hat{p}(1 - \hat{p})/n)}$.

1. With $\hat{p}$ stored as P from above, calculate the margin of error and store in M. $2.576\sqrt{(P(1-P)/2673)}$ **STO▶** M for 0.01216, as in the screen above.

2. Type **P − M ENTER** for 0.0514.

3. Use the last ↑**ENTRY** feature to recall the last line and change subtraction to addition for P + M **ENTER** for 0.0758. Thus, the 99% C.I. = 0.0636 ± 0.0122 = (0.0514, 0.0758).

We are 99% confident that the percentage of adult heterosexuals who had more than one sexual partner in the past year lies between about 5.1% and 7.6% [p. 491].

Hypotheses Testing (p. 490)

Example [p. 491]   The French naturalist Count Buffon tossed a coin 4040 times. He got 2048 heads and 1993 tails. Is this evidence that Buffon's coin was not balanced?

1. $H_0: p = 0.5$ $H_a: p \neq 0.5$.

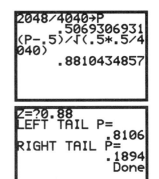

2. $\hat{p} = 2048/4040 = 0.5069$.

3. $z = (\hat{p} - p_0)/\sqrt{(p_0(1 - p_0)/n)}$
   $= (0.5069 - 0.5)/\sqrt{(0.5(1 - 0.5)/4040)} = 0.88$, as in the screen on the right.

4. Running program A1**PTAILS 1**:Z-DIST and entering 0.88 for Z gives a right-tail probability of 0.1894. Because this is a two-tail test, the p-value = 2*0.1894 = 0.3788.

Conclusion: A proportion of heads as far from one-half as Buffon's would happen 38% of the time when a balanced coin is tossed 4040 times. Buffon's result doesn't show that his coin is unbalanced [p. 492].

Choosing the Sample Size

Example [p. 495]   You are planning a sample survey to determine what percentage of voters plan to vote for your candidate for mayor. You want to estimate p with 95% confidence and a margin of error no greater than 3% (or 0.03). How large a sample do you need? The winner's share in all but the most lopsided elections is between 30% and 70% of the vote. So, use the guess $p^* = 0.5$.

$n = (z^*/m)^2 p^*(1 - p^*)$

$\quad = (1.96/0.03)^2(0.5)(1 - 0.5) = 1067.1$, thus 1068, as in the screen on the right.

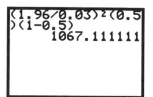

```
(1.96/0.03)²(0.5
)(1-0.5)
        1067.111111
```

COMPARING TWO PROPORTIONS

Confidence Intervals $p_1 - p_2$ (p. 501)

Example [p. 503]   The data below is from a study of the effect of preschool on later use of social services. What is the 95% confidence interval for $p_1 - p_2$?

Population	Population description	Sample size	Need social services	Sample proportion
1	control	n1 = 61	x1 = 49	p̂1 = 0.803
2	preschool	n2 = 62	x2 = 38	p̂2 = 0.613

1. Run program A5**TSAMPS** 2:PROPORTIONS(Z). Make the inputs as in the screen on the right.

2. Press **ENTER** to get the second screen at right. Check the first two lines and ignore the rest until the next example.

3. Press **ENTER** and continue as in the last screen on the right for the confidence interval after entering $z^*=1.96$. You will see $0.190 \pm 0.157 = (0.033, 0.347)$.

Conclusion: We are 95% confident that the percent of people needing social services is somewhere between 3.3% and 34.7% lower among people who attended preschool. The confidence interval is wide because the samples are quite small [p. 504].

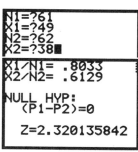

```
N1=?61
X1=?49
N2=?62
X2=?38■
```

```
X1/N1= .8033
X2/N2= .6129

NULL HYP:
   (P1-P2)=0

Z=2.320135842
```

```
Z*=?1.96
(X1/N1-X2/N2)=
       .1903754627
ME=  .157010366
CI FROM/TO
       .0333650967
       .3473858287
```

Significance Tests for $p_1 - p_2$ (p. 504)

Example [p. 505]   The following data is from the Helsinki Heart Study to answer the question of whether gemfibrozil, a drug used to lower blood cholesterol, reduces the risk of heart attack.

Population	Population description	Sample size	Had heart attacks	Sample proportion
1	gemfibrozil	$n1 = 2051$	$x1 = 56$	$\hat{p}1 = 0.0273$
2	control	$n2 = 2030$	$x2 = 84$	$\hat{p}2 = 0.0414$

1. H_o: $p_1 = p_2$ H_a: $p_1 < p_2$.

2. Run program **A5TSAMPS 2**:PROPORTIONS(Z). Note X1 was input on the right as $n_1* \hat{p} = 2051*0.0273$.

Of course, you could have entered 56 but I wanted to show you what to do if you did not have X1.

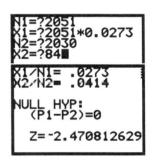

3. Press **ENTER** after inputting X2 to get the screen on the right with the test statistic of z = ⁻2.47.

4. Running program A1**PDIST 3**:Z-DIST gives the screen on the right with the left-tail probability or p-value of 0.0068 < 0.01

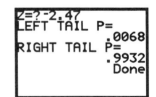

Conclusion: The results are statistically significant at the $\alpha = 0.01$ level. There is strong evidence that gemfibrozil reduced the rate of heart attacks. The large samples in the Helsinki Heart Study helped the study obtain highly significant results [p. 507].

The last two examples could be done without the program **A5TSAMPS2**:PROPORTIONS(Z), but by using the program one is less likely to make a computational error.

CHAPTER 8
Inference for Two-Way Tables

In this chapter we cover the Chi-Square Test for two cases, one that tests several proportions and one that tests whether the row and column variables are related in any two-way table. We will use the program **A6TWTAB** introduced at the end of Chapter 2 for the investigation of the relationship between categorical variables in a table by calculating percentages.

H_o: $p_1 = p_2 = p_3$

Example [p. 522]   The counts of the subjects who avoided relapse and those that relapsed into cocaine use during a study that used medication to fight depression in 72 chronic cocaine users who broke their drug habit are given in the table below.

Group	Treatment	No relapse	Relapsed	Total
1	Desipramine	14	10	24
2	Lithium	6	18	24
3	Placebo	4	20	24
	Total	24	48	72

The expected values are given in the table below. Because 48 out of 72 or two-thirds of the subjects relapsed, it is expected that, of the 24 on each treatment, 16 or two-thirds will relapse if the hypothesis is true [p. 526].

Expected counts		
No relapse	Relapsed	Total
8	16	24
8	16	24
8	16	24
24	48	72

The Components of Chi-Square and the Chi-Square Statistic

1. Enter the observed count in L1 and the corresponding expected value in L2.

2. With L3 highlighted, enter $(L1–L2)^2/L2$, as shown in the first display screen below.

3. Press **ENTER** to fill L3 with the components of chi-square, as shown in the screen in the background below and in the text [p. 530].

L1	L2	L3		L3
14	8	------		4.5
6	8			.5
4	8			2
10	16			2.25
18	16			.25
20	16			1

$L_3 = (L_1 - L_2)^2 / L_2$

4. Return to the home screen with ↑QUIT and sum L3, then **ENTER**. (Note: sum under ↑LIST [MATH] 5:sum) for the Chi-Square Statistic = 10.5 as at the right).

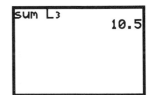

```
sum L3
            10.5
```

Program A4TWTAB

The above calculations are easily handled with this program, first introduced in Chapter 2, where it was used to calculate percentages.

1. Enter the observed counts in Matrix [D].
(a) Press **MATRX** [EDIT] **4**:[D].
(b) Type **3 ENTER 2 ENTER**, as in 3 rows and 2 columns in the table, to get the first screen on the right. Note that if you had a two-digit number that you typed over in this step you have to delete the second digit with the **DEL** key. If you have numbers other than zeros in your matrix, it does not matter because you will type over them.

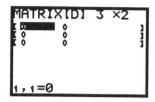

(c) Type **14 ENTER 10 ENTER** for the first row and the cursor advances to the second row. Typing **6 ENTER 18 ENTER 4 ENTER 20 ENTER** completes the data entry. Note that the cursor control keys can be used to move around the editor to check the values and make corrections.

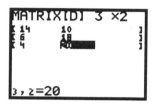

2. (a) Run program A6TWTAB and continue until the screen on the right appears. Press **2**, and after some calculations the screen below appears, with the chi-square statistic = 10.5, as above, but also with the degrees of freedom given (D. F. =2).

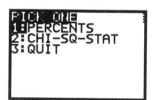

CHI-SQ=
            10.5
D.F.=
            2

(b) Press **ENTER** to show the first screen below and the three selections from that screen brings up the others. Note that the values in these screens are the same as those in L1, L2, and L3 in the original calculations and agree with the text and the Minitab output [p. 530].

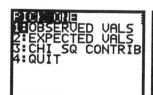
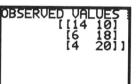
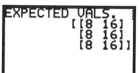

3. Rerun the program but this time pick **1:PERCENTS** and then **3:ROW PERCENTS** to get the screen on the right and as discussed in the text [p. 531], where "desipramine (the first row) has a much higher success rate than either lithium or a placebo" (42% had relapses compared to 75% and 83%).

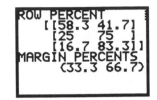

The Chi-Square Distributions (p. 532)

Run program A1**PDIST** 5:CHISQ-DIST and input 10.5 for CHISQ and 2 for D.F. for the p-value = 0.0052, as shown on the right and in the text [p. 530].

Conclusion: There are significant differences in the proportions above. They are not all two-thirds or 66.7% relapses.

H_0: THERE IS NO RELATIONSHIP BETWEEN MARITAL STATUS AND JOB GRADE

Example (p. 535)

		Single	Married	Divorced	Widowed
			Marital status		
Job grade	1	58	874	15	8
	2	222	3927	70	20
	3	50	2396	34	10
	4	7	533	7	4

1. Put the data in Matrix [D] as shown below left.

2. Run program A6**TWTAB** for PERCENTS for the column percentages below right

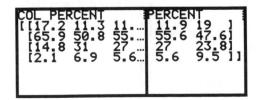

and in the text [p. 538]. "We see at once that smaller percentages of single men (first column) have jobs in the higher grades 3 and 4."

3. (a) Rerunning the program again but for CHI-SQ-STAT alerts us to the fact that we may not meet the cell counts requirement for the chi-square test [p. 539] and that the expected values in the last column (Widowed) are less than five for both the first and fourth rows (Job grade).

(b) This is verified by continuing the program output (the screen below right) with values of 4.87 and 2.81. As pointed out in the text, under the cell count requirement box [p. 540], our example "easily passes this test. All expected counts are greater than 1, and only 2 out of 16 (12.5% ≤ 20%) are less than 5"[p. 540].

```
SMALL EXP FREQ    COL
IN ROW                        4    SMALL EXP FREQ
COL                  1    IN ROW                4
                     4    COL
                          COL                   4
                                                4
```

```
EXPECTED VALS.    EXPECTED VALS.
[[39.08  896.44...    14.61  4.87 ]
 [173.47 3979.0...:5  64.86 21.62]
 [101.9  2337.3...:   38.1  12.7 ]
 [22.55  517.21...:   8.43  2.81 ]]
```

(c) A very large chi-square statistic (67.397) is shown below left.
(d) The chi-square components below right also indicate that the largest contributor to the chi-square statistic is in the first column.

```
CHI-SQ=
               67.397
D.F.=
                    9
```

```
CHI SQ CONTRIBUT CONTRIBUT
[[9.16  .56  .0... .01 2.01]
 [13.58 .68  .4... .41 .12 ]
 [26.43 1.47 .4... .44 .57 ]
 [10.72 .48  .2... .24 .5  ]]
```

4. Run program A1PDIST 5:CHISQ-DIST, and with the appropriate inputs, the p-value is quite small (0.0000+), so there is no hesitation in rejecting the null hypothesis and concluding that there is a significant association between marital status and job grade.

```
CHISQ=?67.397
D.F.=?9
LEFT TAIL P=
            0.9999+
RIGHT TAIL P=
            0.0000+
            Done
```

Of course, this association between marital status and job grade does not show that being single causes lower-grade jobs. The explanation might be as simple as the fact that single men tend to be younger and so have not yet advanced to higher grades [p. 539].

CHAPTER 9

One-Way Analysis of Variance: Comparing Several Means

Program A8**ANOVA** will be used for doing hypothesis tests comparing several means. Both raw data and summary values will be used for input.

Example [p. 556] City gas mileage (mpg) is given for 1994 car models. Do small cars really have better gas mileage? (See the table on the following page.)

1. Compare the different-size cars by putting the data in L1, L2, and L3 and defining the plots appropriately (as in Chapter 1) to draw the following histograms, all with the same WINDOW as in the screen below left.

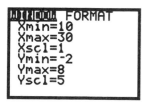
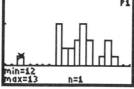
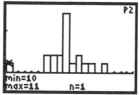
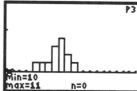

Notice as you go right to larger cars, the distributions shift left and there are two outliers. The shifting to the left indicates that, on average, the larger the car, the lower the gas mileage. The two outliers are boxed in the table. "The Jaguar XJ12 and the Rolls-Royce Silver Spur are responsible. These are rather exotic cars, so we will omit these outliers from all further analyses [p. 558].

2. Go to the Stat editor and use **DEL** to delete the outliers from L1 and L2. Use 1-VarStats (under **STAT** [CALC]) on each list for the following:

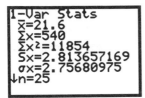

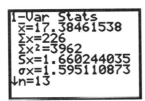

3. Turning on a different box plot for each list, there is a shift to the left of the distributions again (see below) also verified by the means getting smaller (21.6, 19.32, 17.38). We also notice the spread is decreasing, which is also verified above with the standard deviations decreasing (2.81, 2.31, 1.66).

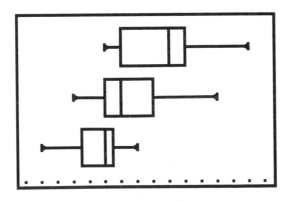

L1	L2	L3
Compact	Midsize	Large
19	16	19
20	25	19
18	19	17
18	16	16
18	19	18
22	22	20
25	20	18
22	19	18
23	21	18
22	17	17
23	17	15
12	18	14
23	18	17
26	23	
19	19	
18	19	
18	10	
20	19	
27	21	
21	19	
22		
26		
22		
26		
21		
21		

"The difference among the means is not large. Are they statistically significant?" [p. 558].

THE ANALYSIS OF VARIANCE F TEST AND PROGRAM A8ANOVA

Ho: $\mu 1 = \mu 2 = \mu 3$

1. Run program A8**ANOVA** to get the screen on the right. Press **1:** ONE-WAY and continue until you are given the choice of the next screen. Press **2** because you will use summary statistics as input. Matrix [D] will be used in the next example.

2. Enter **3** at the HOW MANY SAMPLE... prompt, then enter the summary statistics (mean, standard deviation and sample size) at the appropriate prompt, as shown in the screens on the right.

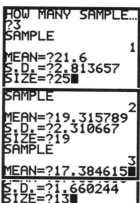

3. Press **ENTER** to bring up the screen below with F = 13.62, which agrees with the Minitab output of the text [p. 560]. (Note that we needed to enter more decimals than we would usually carry for inputs because we want to compare our answers to the Minitab output that was based on all the data with many decimals carried until the final display below.)

```
        DF   SS
FAC   2   160.95819
ERR  54   319.18217
45
    F=13.61564567
    SP=2.431209783
```

The one concession made because of the smaller screen for the TI-82, compared to a Minitab display, is that TOTAL DF and SS and the MS values are not given. However, these are all easily obtained from the values given e.g., 160.95819/2 = 80.48 for MS FACTOR. The p-value is below.

4. Press **ENTER** after the above to either QUIT or continue for the individual confidence intervals based on the pooled standard deviation or SP. Choose the latter to get the following screen.

To obtain the 95% confidence intervals, look up D.F. = 54 in Table C and settle for the conservative D.F. = 50 with an upper tail probability of 0.025 for the critical t value or t* = 2.009.

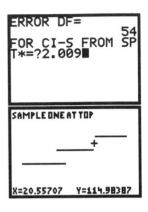

5. Type this in at the prompt, press **ENTER** and the display on the right is obtained. Press **ENTER** for "Done," then **GRAPH**, then with the cursor control keys move the cross hair to the position shown with X = 20.557 (ignore Y as it has no meaning in this analysis). Sample one's confidence interval is above 20.6 mpg and does not overlap with the intervals for the larger size cars.

6. ♠ **QUIT** returns to the home screen.

7. Run program A1**PDIST 6**: F-DIST and enter **13.62** at the F prompt, **2** (D.F. factors) for D.F. NUM., and **54** (D.F. error) for D.F. DEN. for a very small p-value = 0.0000+ as shown in the screen on the right.

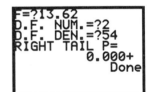

Conclusion: "There is strong evidence (p < 0.001) that the means are not all equal, and the most important difference among the means is that compact cars have better gas mileage than midsize and large cars." [p. 561]

Program A8ANOVA Using Raw Data in Matrix [D]

Example [p. 572] Which colors of sticky boards attract insects best? See the data in the table on the right.

Board color	Insects trapped					
(1) Blue	16	11	20	21	14	7
(2) Green	37	32	20	29	37	32
(3) White	21	12	14	17	13	20
(4) Yellow	45	59	48	46	38	47

At the right is how the data will be entered into [D] with all the data for board color 1 (blue) first, followed by the data for board color 2 and so on.

1. Enter the data in Matrix [D].
(a) Press **MATRX** [EDIT] **4:[D]**.

(b) Type **24 ENTER 2 ENTER**, because there are 24 values or rows and 2 columns (the first column for the data and second column for color number; these must be consecutive integers starting with 1), to get the screen below. Note if you had two digits where we now have 2, you would have to delete the second digit with the **DEL** key. If you have other than zeros in your matrix, it does not matter because you will type over them.

16	1
11	1
20	1
21	1
14	1
7	1
37	2
32	2
20	2
29	2
37	2
32	2
21	3
12	3
14	3
17	3
13	3
20	3
45	4
59	4
48	4
46	4
38	4
47	4

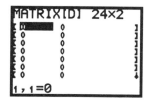

(c) Type **16 ENTER 1 ENTER** for the first row, and the cursor advances to the second row. Type **11 ENTER 1 ENTER 20 ENTER 1 ENTER ...47 ENTER 4 ENTER** so your matrix is filled, as on the right. Note that the cursor control keys can be used to move around the editor to check the values and make corrections. You could also have entered your data by columns with these keys.

(d) Press ↟**QUIT** to return to the home screen.

(2) Run program A8**ANOVA 1**: ONE-WAY but this time for DATA INPUT WITH select **1:**DATA MAT [D] for the following output, which includes the means, standard deviations, and sample size. This agrees with the Minitab output of the text [p. 574].

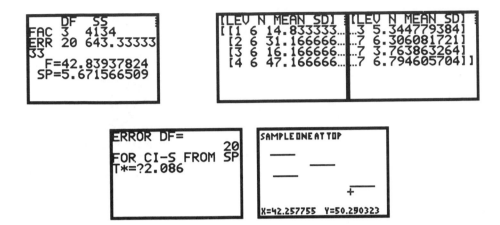

Conclusion: We see that sample four (yellow boards) has the largest number of insects. "Yellow boards appear best at attracting leaf beetles" [p. 573].

Pooled Standard Deviation and Confidence Intervals (p. 581)

Example [p. 581] The calculation for the 95% confidence interval for the mean count of insects trapped by yellow boards using the sample size of 6 and mean of 47.167 for LEV 4 (from the screen on the upper right in step 2 above) and SP = 5.672 and degrees of freedom, D.F. = 20 (from the ANOVA screen on the upper left in step 2) and a critical t value of 2.086 from Table C of the text and above.

$$\overline{x}_4 \pm t^* \, SP/\sqrt{n_4} = 47.167 \pm (2.086)5.672/\sqrt{6}$$
$$= 47.167 \pm 4.830$$
$$= 42.34 \text{ to } 52.00$$

These calculations are shown in the screen on the right and in the graphics screen above. The X value of the cross hair above is at 42.26, the closest pixel to 42.34.

```
2.086*5.672/√6→M
        4.830308857
47.167-M
        42.33669114
47.167+M
        51.99730886
```

CHAPTER 10

Inference for Regression

In Chapter 2 we dealt with the Sanchez's natural gas consumption data (duplicated below). We looked at the scatter plot, the correlation coefficient and the least-squares regression line to predict the gas consumption for a month with a given value for degree-days. "Now we want to do tests and confidence intervals in this setting" [p. 590].

We will follow this data along for a number of examples in the text using the program A4**LINREG** for most of the plotting and calculations.

REGRESSION LINE AND CORRELATION

Example 10.2 [p. 591]

1. Put degree-days in L1 and gas usage in L2.

2. Run program A4**LINREG** and make the first two selections for the PLOT and for a, b, r^2, etc., as was done in Chapter 2, for the two rightmost screens on the following page.

Month	Deg-(L1) days	Gas (L2) (100 cu ft)
Nov.	24	6.3
Dec.	51	10.9
Jan.	43	8.9
Feb.	33	7.5
Mar.	26	5.3
Apr.	13	4
May	4	1.7
June	0	1.2
July	0	1.2
Aug.	1	1.2
Sept.	6	2.1
Oct.	12	3.1
Nov.	30	6.4
Dec.	32	7.2
Jan.	52	11
Feb.	30	6.9

We see that $\hat{y} = a + bx = 1.08921 + 0.1890x$ and $r = 0.9953$ and $r^2 = 0.9906$.

The first screen below informs us that the fits or yhats are stored in L3 and the residuals are stored in L4. This information is used in the next section.

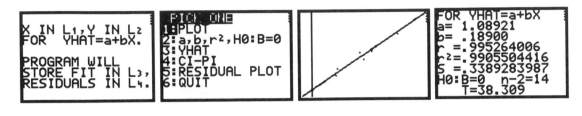

STANDARD ERROR ABOUT THE REGRESSION LINE AND RESIDUALS

Example 10.4 [p. 595]

1. Continuing with the chapter example, we want to predict y for x = 24 so we press **3:YHAT**, and at the X=? prompt, type **24** then **ENTER** for the predicted value of 5.6252.

```
INPUT X=?24
YHAT=5.625185735
```

2. The value from the original data for (Nov.) is
x = 24, y = 6.3, so the residual =
$y - \hat{y} = 6.3 - 5.6252 = 0.6748$.

Note this is just the value found in the first row of the spreadsheet in L4 after running A4LINREG (see the screen on the right).

L2	L3	L4
6.3	5.6252	.67481
10.9	10.728	.17184
8.9	9.2162	-.3162
7.5	7.3262	.17382
5.3	6.0032	-.7032
4	3.5462	.4538
1.7	1.8452	-.1452

L4(1)=.67481426...

3. Note that s^2 equals the sum of the residuals squared divided by n−2 or 14, and s is the square root of s^2. Because the residuals are in L4, you can use the sum function under ↑LIST [MATH], as in the second screen on the right, to calculate s = 0.3389.

```
√(sum (L4²)/14)
          .3389283987
```

The above was not necessary because s was on one of the output screens from program A4LINREG as capital S. (See the sixth line of the screen on the upper right from the previous section.)

TESTING THE HYPOTHESIS OF NO LINEAR RELATIONSHIP H_O: $\beta = 0$

Example 10.6 [p. 601]

1. The last line of the upper right screen on page 88 shows that the test statistic of t = 38.309 is very large.

2. Running program A1**PDIST** 4:T-DIST with the input of T=38.309 and D.F.=14 gives a very small p-value, as in the screen on the right. Reject the hypotheses of no slope and conclude that there is a significant positive slope and significant positive linear correlation.

```
T=?38.309
D.F.=?14
LEFT TAIL P=
          0.9999+
RIGHT TAIL P=
          0.0000+
          Done
```

CONFIDENCE INTERVALS FOR THE REGRESSION SLOPE

Example 10.5 [p. 599]

1. The t statistic above was calculated as T = $(b - 0)/SE_b$ thus $SE_b = b/T = 0.1890/38.309 = 0.004934$.

```
0.1890/38.309
        .0049335665
2.145*Ans→M
        .0105825002
0.1890-M
        .1784174998
0.1890+M
```

2. The 95% confidence interval = b ± t*SE_b = 0.1890 ± 2.145*0.004934 for an interval from 0.1784 to 0.1996, as on the right and in the text, with 2.145 being the critical t for 14 D.F. from Table C of the text.

```
0.1890+M
        .1995825002
```

INFERENCE ABOUT PREDICTIONS: CONFIDENCE AND PREDICTION INTERVALS

Examples 10.8 [p. 605] **and 10.9** [p. 607]   To predict gas consumption at 20 degree-days, substitute 20 for x in the regression equation. To obtain the 95% confidence and prediction intervals:

1. Run program A4**LINREG** and from the option screen on the left of the following page select **4: CI-PI** for the second screen.

2. Input the critical t value for the confidence and prediction levels of interest (T* = 2.145 for the 95% level as above) and the desired x (X = 20), and you

obtain the 95% confidence interval from 4.6858 to 5.0526 and the 95% prediction interval from 4.1194 to 5.6190, as below right and in the text [p. 608].

Note that the last line of the screen gives YHAT of about 4.869 hundred cubic feet of gas per day.

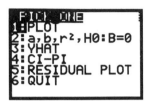
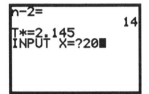
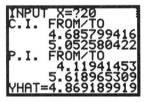

CHECKING THE REGRESSION ASSUMPTIONS

Example [p. 601]   How well do golfers' scores in the first round of a two-round tournament predict their scores in the second round? The data are given below.

Golfer	1	2	3	4	5	6	7	8	9	10	11	12
Round 1 (L1)	89	90	87	95	86	81	102	105	83	88	91	79
Round 2 (L2)	94	85	89	89	81	76	107	89	87	91	88	80

1. With the data in L1 and L2 as indicated in the table above, run program A4LINREG for the output below that agrees with the text [pp. 601–603].

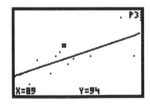
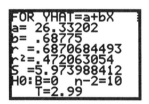

2. Select **5** from the option screen on the left on the following page for the residual plot that gives us the plot below center in trace mode, where the cursor key was used to highlight the seventh golfer with a residual of 10.51779, and a first-round score of 102 in L1 and a second-round score of 107 in L2. Note that this is the seventh value in L4 (rightmost screen), just as in the text [p. 612]. Note also that after the program is run the residuals are

stored in L4. This is a reasonable scatter for the residuals as explained in the text [p. 611].

 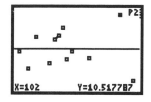 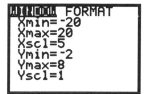

3. To check the distribution of the residuals for signs of strong nonnormality, a stem plot was given in the text [p. 612]. A similar distribution is obtained by plotting a histogram of the residuals in L4 with the WINDOW set-up and plot defined as follows:

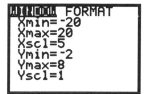  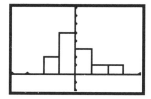

"The distribution is quite symmetric. Inference using the assumption that these vary normally will give approximately correct results" [p. 612].

So it is that we end the last chapter of this guide with some of the procedures that we started with in Chapters 1 and 2. They have served us well.

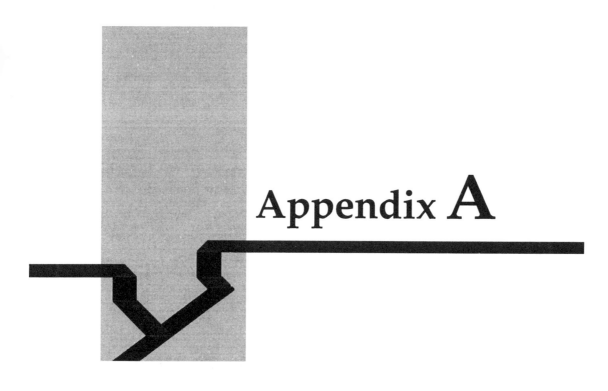

Appendix A

LOADING DATA AND PROGRAMS

A disk to accompany this guide is available from the publisher in both DOS/Windows and Macintosh formats and contains the following:

1. A group of statistical programs, with the file name MSTATPAK.82G, that are used throughout this guide and complement the built-in routines of the T1-82. The "Introduction to the TI-82" at the beginning of this guide briefly describes each program and provides the page where you can find an example of how to use the program.

2. The data sets mentioned in the Preface are given in ASCII format. The file names follow the usual convention: tab2-1.dat for Table 2.1, ex2-4.dat for Exercise 2.4, em2-6 for Example 2.6 and guid2.dat for Instructors Guide data set 2. See the README file or the Instructors Guide for more details.

Your instructor will probably load the programs of MSTATPAK into your TI-82 early in the course and perhaps transfer some of the larger data sets later, when you are planning to work with them. You may want to investigate the last section on sharing data or reloading a program with a classmate's TI-82.

Loading MSTATPAK.82G and/or DATA from a Computer to a TI-82 (Your TI-Graph Link Guidebook also gives instructions.)

1. Connect your TI-82 to your computer with the TI-GRAPH LINK cable.

2. Press ⬆ **LINK** on your TI-82, then ▶ to display RECEIVE. Press **1** to select Receive and have the message "Waiting...." displayed.

3. (a) For MSTATPAK.82G: On the computer with the TI-GRAPH LINK82 software, open the SEND menu and select Group. Move along the file names for MSTATPAK and SELECT on the PC or click to ADD on the Macintosh. Then press Xmit on the PC or click on DONE and then SEND on the Macintosh.
(b) To load data from an ASCII file, before opening the Send menu, you must open the Utilities menu and select "Import ASCII file," then import to the desired List or Matrix. See TI-Graph Link Guidebook for full details.

Loading PROGRAMS and/or DATA from one TI-82 to another

Note that when MSTATPAK.82G was transferred from the computer to a TI-82 it was rearranged into individual programs. The complete list of programs is given in the "Introduction to the TI-82."

This procedure is similar to the section above, but the TI-82's are now connected together by the cable that came with these units. Use ⬆ **LINK** on both calculators, one set to RECEIVE and the other to TRANSMIT. See your TI-82 Guidebook (Chapter 16) for complete instructions.

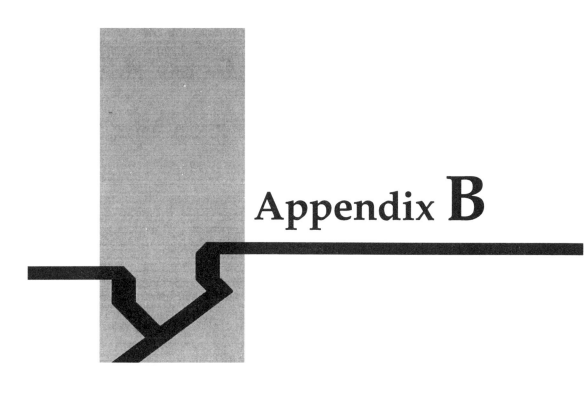

Appendix B

THE CALCULATOR BASED LABORATORY (CBL) SYSTEM FOR DATA ACQUISITION AND ANALYSIS WITH THE TI-82

Much data is taken dynamically (in real time) and electronically for analysis. We can now do this data gathering with little technical background using a handheld unit. The CBL system from Texas Instruments comes with a workbook that has detailed instructions and listings of the necessary programs to do many experiments. As more people become familiar with this system the number of experiments should grow accordingly.

The first experiment in this appendix is modified from the workbook and uses the temperature probe that comes with the CBL unit and was selected because it is easy to set up and carry out, primarily as a demonstration. The second experiment was selected because it shows experimental design considerations and execution and permits full-class involvement. It uses the force sensor, mentioned in the workbook, that must be purchased separately from the basic CBL unit.

NEWTON'S LAW OF COOLING

See Experiment P3 in the CBL System Experiment Workbook. This experiment works well after reading the linear regression discussions in Chapter 2 to show that one can fit other than a straight line to data by the fitting of an exponential curve.

A hot cup of coffee cools because the surrounding air is at a lower temperature. The greater the temperature difference, the more rapid the cooling. As the coffee approaches room temperature, it cools very slowly and remains lukewarm for a long period. Its temperature is lowering at a decreasing rate instead of the constant rate that would be indicated as a straight line on a plot of coffee temperature versus time. To show this, do the following:

1. Take some boiling water and put it in a thermos for safe and easy transport.

2. Connect the TI-82 to the CBL as if it were another TI-82. Connect the temperature probe to CH2 on the top edge of the CBL unit.

3. (a) Run Program COOLTEMP from the workbook with the temperature probe at room temperature. To avoid conduction and evaporation effects on the temperature probe, do not place the probe directly on the table top or expose it to any drafts.

(b) The program runs for about 1.5 minutes, but after about 30 seconds put the probe into the hot water and watch the temperature rise (to about 85°C in our example). See the display screen on the right.

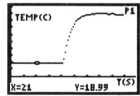

(c) After the program stops running, press **TRACE** and read the room temperature (18.99°C) as shown in the previous screen.

4. Lift the probe from the water, flick off the last drop of water, and run the program COOLTEMP for the cooling curve at the right. (Keep the probe out of drafts and do not let it touch the table top.) Notice how the temperature levels off to the room temperature.

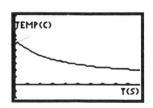

5. The rate of change of the temperature is proportional to the difference between the temperature of the liquid and the room temperature, so remember to subtract room temperature from each of the liquid readings stored by the program in L4.

(a) From the home screen take L4 – 18.99 **STO▶**L4.

(b) Press **GRAPH** to get the screen below on the left. When the room temperature is subtracted the curve approaches zero.

(c) Press **ZOOM 9:**StatZoom and the data fills more of the screen, as in the screen below on the right.

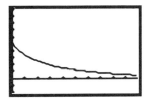
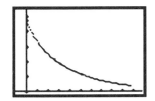

6. (a) The line is not straight and if you press **STAT**[CALC] **9:** LinReg(ax+b) **L2, L4 ENTER** (the program stores the time in L2 and the temperature in L4), you get r = –0.927039487, as in the screen on the right with $r^2 = 0.8594$.

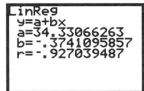

(b) Theory predicts exponential cooling, so **STAT**[CALC] **A:**ExpReg **L2, L4 ENTER** gives you r = –0.9980841837 or $r^2 = 0.9962$, as in the screen on the right. Note that r is the linear correlation coefficient from fitting the data to the straight line ln y = ln a + (ln b)*x.

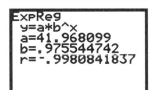

(c) Press **Y=** and set Y1 = 41.968*0.9755^X and **GRAPH** for the next screen on the right, which has this curve fitting the data very well except for the first data points. Note that 41.968* 0.9755^X = 41.968*e^(ln0.9755*X) = 41.968*e^(⁻0.0248X) for those who like their exponential curves to have a base.

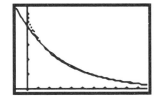

HAND STRENGTH

A. Is the right hand generally stronger than the left in right-handed people? You can crudely measure hand strength by placing a bathroom scale on a shelf with the end protruding, then squeezing the scale with the thumb above and four fingers below. The reading of the scale shows the force exerted. Describe the design of a matched-pairs experiment to compare the strength of the right and left hands, using 14 right-handed people as subjects (7 male and 7 female).

Note: The above was modified somewhat from text Exercise 3.42 [p. 216].

B. Are males stronger than females (hand strength for right-handers)?

Apparatus

The advantage of using the apparatus shown below over the bathroom scale is that everyone can follow along with the readings with the classroom display unit for the TI-82. The apparatus is easily made with wood and nails and one bolt for a pivot. It is basically a seesaw with about 2 inches (of a 1.5" by 1" by 32" plank) from the pivot to the thumb and about 30 inches to the force sensor. This is to protect the sensor, which would not be able to measure a force as large as needed and is easily handled by a scale. If you do not have a force sensor it is still worth doing these exercises with a bathroom scale.

Force Sensor

Taking and Recording Data

1. Connect the TI-82 to the CBL as if it were another TI-82. Connect the force sensor to channel one (CH1) on the top edge of the CBL unit.

2. Make sure the CBL and the TI-82 are turned on. Start the program FORCERT listed in the CBL workbook (with the program change of 15 for Ymax in line 3 and add line 22 as in the program printout on page 101).

3. When you are ready to start collecting data (see Experimental Design below), place a thumb on the end of the seesaw and the fingers under the board protruding over the edge of the desk. Press **ENTER** on the TI-82 to start the force graph. Apply pressure evenly, trying not to jerk start, while watching the plotting of the sensor output. Try to hold the maximum pressure for a few moments and then evenly release the pressure. Repeat with the other hand for a plot like that in the screen on the right.

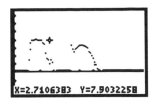

X=2.7106383 Y=7.9032258

Note: This procedure is best demonstrated so that the subjects get an idea of the timing required to get outputs for both hands on the same graph. If there is a difficulty, just start over.

4. Use the cursor control key to read and then record the high points for the right (7.90 as in the screen) and left hands. An agreement that at least two pixels at the same level must be on works well to eliminate the one-pixel impulse forces.

Experimental Design

1. Randomly select 7 males and 7 females from the class. They will perform the experiment in the order that they are selected. Let the males go first.

2. Flip a coin for each participant. If the toss is heads, subjects will use the right hand first. If tails, the left hand is used first.

Example Data and Analysis

		Male			Female	
Order	First hand	Right	Left	First hand	Right	Left
1	L	7.9	6.61	R	6.45	6.77
2	L	5	5.32	R	4.19	2.9
3	L	6.29	4.68	L	3.55	4.52
4	R	8.87	6.94	R	4.52	3.55
5	L	6	4.68	R	3.23	2.26
6	R	5	4.68	R	4.52	3.55
7	L	7.9	8.55	L	6	3.64

A. Is the right hand generally stronger than the left in right-handed people?

1. $H_0: \mu_R - \mu_L = 0$ $H_a: \mu_R - \mu_L > 0$

2. (a) Put data in L1 and L2 and the difference in L3 = L1 – L2.
(b) 1-VarStats L3 gives the mean difference $\bar{x}$ = 0.7693 and the standard deviation of the differences Sx = 1.006.
(c) t = (0.7693–0)/(1.006/√14) = 2.861.

3. Using program A2PTAILS 3:T-DIST with t= 2.861 and D.F. = 13, there is a p-value of approximately 0.0067 and the conclusion that the right hand is significantly stronger than the left.

L1(Right)	L2(Left)	L3=L1-L2	L4=(L1+L2)/2	L5(x)	L6(y)
7.9	6.61	1.29	7.255	7.255	6.61
5	5.32	-0.32	5.16	5.16	3.545
6.29	4.68	1.61	5.485	5.485	4.035
8.87	6.94	1.93	7.905	7.905	4.035
6	4.68	1.32	5.34	5.34	2.745
5	4.68	0.32	4.84	4.84	4.035
7.9	8.55	-0.65	8.225	8.225	4.82
6.45	6.77	-0.32	6.61		
4.19	2.9	1.29	3.545		
3.55	4.52	-0.97	4.035		
4.52	3.55	0.97	4.035		
3.23	2.26	0.97	2.745		
4.52	3.55	0.97	4.035		
6	3.64	2.36	4.82		

B. Are males stronger than females (hand strength for right-handers)?

Consider that the samples of males and females are independent samples of a larger population of all students.

1. Ho: $\mu_M = \mu_F$ $H_a: \mu_M > \mu_F$

2. (a) With the average force measurements of right and left hands for each subject in L4 = (L1 + L2)/2.
(b) Put the first seven values, males, in L5, or x, and the female data in L6, or y.

(c) Use 2-VarStats L5, L6 for $\bar{x}$ = 6.316 and Sx = 1.426 $\bar{y}$ = 4.261 and Sy = 1.211.
(d) Run program A5**TSAMPS** 1:MEANS(T-TEST) with the input as indicated in the screen below on the left and the output as indicated in the screen below on the right.

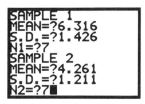

3. Using program A1**PDIST** 4:T-DIST with t = 2.906 and D.F. = 11, there is a p-value of approximately 0.007 (see the screen on the right) and the conclusion that male hand strength is significantly stronger than females.

```
T=?2.906
D.F.=?11
LEFT TAIL P=
            .9929
RIGHT TAIL P=
            .0071
          Done
```

Note: If a matched pairs design is not used to test the hypothesis that the right hand is stronger than the left and L1 and L2 are considered as two independent samples, the inputs and outputs of the first two screens on the right (5.673 – 4.904 = 0.769) would be obtained and there would be insufficient evidence to conclude that there is any significant difference in hand strength (see the last screen on the right).

```
SAMPLE 1
MEAN=?5.673
S.D.=?1.700
N1=?14
SAMPLE 2
MEAN=?4.904
S.D.=?1.767
N2=?14
```

```
H0:(MU1-MU2)=
            0
TEST STAT T=
          1.173
  D.F.=25.96
POOLED T=
          1.173
  D.F.=26
```

```
T=?1.173
D.F.=?25
LEFT TAIL P=
            .8741
RIGHT TAIL P=
            .1259
          Done
```

Program FORCERT (modified)

```
PlotsOff :FnOff :AxesOn
0→Xmin:99→Xmax
⁻5→Ymin: 15→Ymax: 10→Xscl :2→Yscl
ClrList L2,L4
{1,0}→L1
Send(L1)
ClrHome
{4,1,1,1,9.8,⁻9.8}→L1
Send(L1)
```

```
{1,1,1,0,0,1}→L1:Send(L1)
Disp "PRESS ENTER TO"
Disp "START FORCE":Disp
"GRAPH.":Pause
{3,.1,⁻1,0}→L1
Send(L1)
99→dim L4
ClrDraw
Text(4,1,"FORCE(N)")
Text(51,81,"T(S)")
For(I,1,99,1)
Get(L4(I))
⁻1*L4(I)→L4(I)
Pt-On(I,L4(I))
End
seq(N,N,0,9.8,0.1)→L2
9.8→Xmax
1→Xscl
Plot1(Scatter,L2,L4,.)
DispGraph
Text(4,1,"FORCE(N)")
Text(51,81,"T(S)")
Stop
```

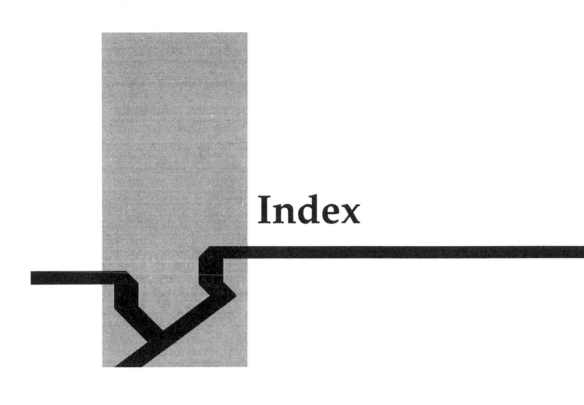

Index

TI-82 Quick Reference

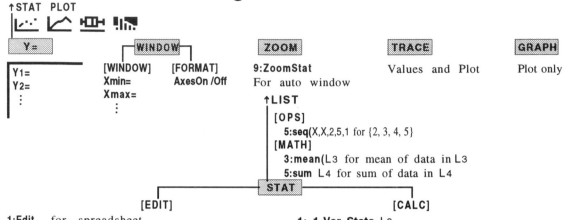

↑STAT PLOT

| Y= | WINDOW | ZOOM | TRACE | GRAPH |

Y=

Y1=
Y2=
⋮

[WINDOW]   **[FORMAT]**
Xmin=   AxesOn /Off
Xmax=
⋮

ZOOM
9:ZoomStat
For auto window

TRACE
Values and Plot

GRAPH
Plot only

↑LIST

[OPS]
  5:seq(X,X,2,5,1 for {2, 3, 4, 5}
[MATH]
  3:mean(L3 for mean of data in L3
  5:sum L4 for sum of data in L4

STAT

[EDIT]

1:Edit... for spreadsheet

L1	L2	L3
------	------	------

2:SortA(L2 Sorts data in L2 from low to high.

3:SortD(L3 Sorts data in L3 from high to low.

4:ClrList L2,L4 Clears data from L2 and L4.

[CALC]

1: 1-Var Stats L3
  For x̄, Sx, Min, Q1, Med, Q3, Max of data in L3.
1: 1-Var Stats L4,L5
  Output as above for data values in L4 with
  frequencies in L5 (max of 99, if 150 values
  of 6, list 6 twice in L4 and a 99 and 51 in L5.
2: 2-Var Stats L2,L3
  For x̄, Sx, ȳ, Sy, n...with equal sample sizes
  and with X values in L2 and Y values in L3.
9: LinReg(a+bx) L3,L4
  For a, b, and r with X values in L3 and Y in L4.

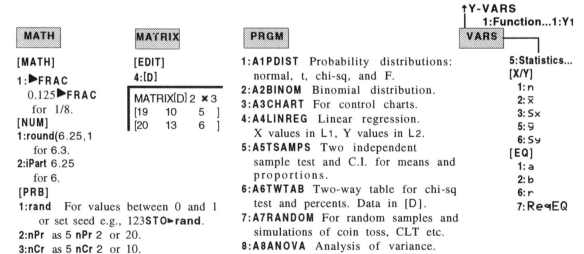

↑Y-VARS
  1:Function...1:Y1

| MATH | MATRIX | PRGM | VARS |

[MATH]
1:▶FRAC
  0.125▶FRAC
  for 1/8.
[NUM]
1:round(6.25,1
  for 6.3.
2:iPart 6.25
  for 6.
[PRB]
1:rand For values between 0 and 1
  or set seed e.g., 123STO▶rand.
2:nPr as 5 nPr 2 or 20.
3:nCr as 5 nCr 2 or 10.
4:! as 5! or 120.

[EDIT]
4:[D]

MATRIX[D] 2 ✕ 3		
[19	10	5]
[20	13	6]

PRGM

1:A1PDIST Probability distributions:
  normal, t, chi-sq, and F.
2:A2BINOM Binomial distribution.
3:A3CHART For control charts.
4:A4LINREG Linear regression.
  X values in L1, Y values in L2.
5:A5TSAMPS Two independent
  sample test and C.I. for means and
  proportions.
6:A6TWTAB Two-way table for chi-sq
  test and percents. Data in [D].
7:A7RANDOM For random samples and
  simulations of coin toss, CLT etc.
8:A8ANOVA Analysis of variance.

VARS

5:Statistics...
[X/Y]
1: n
2: x̄
3: Sx
5: ȳ
6: Sy
[EQ]
1: a
2: b
6: r
7: RegEQ